THE PACEMAKER CHANNELS
OF THE
HEART
From Reductionism to Systems Biology

THE PACEMAKER CHANNELS
OF THE
HEART
From Reductionism to Systems Biology

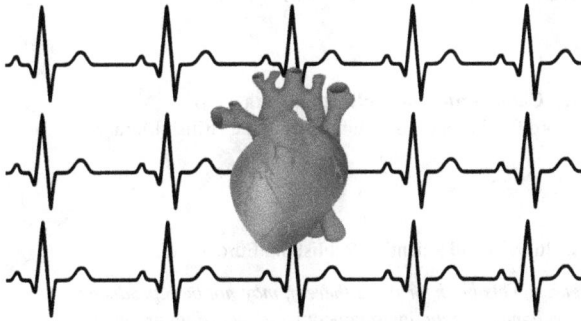

Denis Noble

University of Oxford, UK

Daegu Gyeongbuk Institute of Science and Technology, South Korea

W₢ World Scientific

NEW JERSEY · LONDON · SINGAPORE · BEIJING · SHANGHAI · HONG KONG · TAIPEI · CHENNAI · TOKYO

Published by

World Scientific Publishing Europe Ltd.
57 Shelton Street, Covent Garden, London WC2H 9HE
Head office: 5 Toh Tuck Link, Singapore 596224
USA office: 27 Warren Street, Suite 401-402, Hackensack, NJ 07601

Library of Congress Cataloging-in-Publication Data
Names: Noble, Denis, 1936–   author
Title: The pacemaker channels of the heart : from reductionism to systems biology / Denis Noble.
Description: London : World Scientific Publishing Europe Ltd, [2026] |
    Includes bibliographical references and index.
Identifiers: LCCN 2025031576 | ISBN 9781800618114 hardcover |
    ISBN 9781800618121 ebook | ISBN 9781800618138 ebook other
Subjects: LCSH: Heart--Electric properties--Mathematical models--History |
    Systems biology--History
Classification: LCC QP112.5.E46 N63 2025
LC record available at https://lccn.loc.gov/2025031576

British Library Cataloguing-in-Publication Data
A catalogue record for this book is available from the British Library.

For any available supplementary material, please visit
https://www.worldscientific.com/worldscibooks/10.1142/Q0532#t=suppl

Desk Editors: Murali Appadurai/Gabriel Rawlinson

Typeset by Stallion Press
Email: enquiries@stallionpress.com

About the Author

Denis Noble was the first scientist in 1960 to use mathematical models of the pacemaker rhythm of the heart to show that the rhythm emerges automatically from the interaction between ion channel proteins and the electrical field across the cell membrane. This book tells the story of that discovery and its subsequent development over a period of 65 years. He is a leading proponent of theories of evolution in opposition to gene-centric explanations. He is Emeritus Professor and Director of Computational Physiology at the University of Oxford, and a Distinguished Chair Professor in Biomedical Science and Engineering Interdisciplinary Studies at the Daegu Gyeongbuk Institute of Science and Technology in South Korea.

Contents

Introduction

For all animals that need a circulation to supply nutrients to the cells of the body, a heart is needed to pump those fluids. The heart is therefore one of the first organs to start functioning, even when the embryo is very small. In humans, the heart starts beating at around 28 days, when the embryo is much smaller than a centimetre in size.[1] It is essential that it should do so because molecules can rapidly diffuse only over very short distances, well before there is a functioning nervous system to control the other muscles of the body. So, how does the heart do that all by itself? All our skeletal muscles need the nervous system to make them contract. How the heart can beat spontaneously is what this book is about. The heart must somehow excite itself to start beating spontaneously without any nervous signal, and it continues to do that throughout life. The beat itself is mechanical, a strong muscular pump, but what triggers that forceful contraction is electrical.

When I became a research student in 1958, this was one of the wide-open questions in physiology.[2] How can the heart generate its own electrical stimulus? It was my privilege as a student to discover an answer to how it happens. This book tells the story of that discovery and its many consequences over a career spanning two-thirds of a century.

---

[1]Männer J. (2022). When does the human embryonic heart start beating? A review of contemporary and historical sources of knowledge about the onset of blood circulation in man. *Journal of Cardiovascular Development and Disease* 9(6), p. 187. doi: 10.3390/jcdd9060187. PMID: 35735816; PMCID: PMC9225347.

[2]See Choudhury, R., 2024. *The Beating Heart.* London: Head of Zeus, especially Chapter 8.

My field of research is not usually thought to be readily accessible. Electrophysiology is an area of physiology that most biology and medical students avoid. It requires considerable knowledge of the relevant physics and mathematics, as well as demanding biology. Moreover, in the case of the heart, it has been highly controversial. In writing this book, my first reaction was therefore to doubt whether I could write the story for a general audience.

Controversy: that is what changed my mind. I realised that if I could bring the sometimes fierce scientific and philosophical controversies out into the open and let people see how scientists deal with controversy, there would be a story to tell, even an exciting one. Moreover, I have been involved in and at the centre of much of that controversy for more than 60 years. My first publication describing how electrical rhythm could be generated in the heart was published in *Nature* in 1960. There must be few scientists who have pioneered a field and who are still highly active in the field two-thirds of a century later. I have seen heresies become orthodoxies in my own scientific lifetime. That is a rare privilege.

But it is not just the story of those controversies and their successful resolutions. It is also the story of a personal journey. I myself as a scientist and, to some degree, a philosopher and mathematician have been profoundly affected by the journey. The person who started the journey in 1960 is so foreign to me that, in Chapter 8, I even talk about him in the third person.

So, this story is also that of a personal transformation. Particularly since the publication of *The Music of Life* in 2006 and its subsequent rapid translation into ten foreign languages, starting with the French translation, *La Musique de la Vie*, I am sometimes described as one of the pioneers of systems biology. That is correct in the sense that what I was doing 65 years ago, in discovering the tiny electric currents flowing through potassium channels in the heart and then building the first mathematical model of cardiac rhythm, certainly was a form of systems biology. I was putting together a system of interacting proteins within the context of a complete living cell. Rhythm was shown to arise as an integrative property of that system. It is not a property of the individual molecular components. But, as I explain in the first chapter of this book, this is not how I would have explained it myself at the time. Indeed, I nearly lost the opportunity to use the early mainframe computer I needed precisely because I could not explain it as I would do so today. The guardians of the computing machine, on which time was so precious, were highly sceptical both of my ability to do what I was proposing and whether that proposal itself was coherent.

I was lucky. Not only to have the experimental material to make such a proposal to model heart activity but also in many other respects. University College London, where I was studying in 1960, accommodated an oddball student with open arms. I was allowed to roam around the maths lectures, the philosophy seminars, computer programming courses and much else, while pursuing my PhD in physiology under my supervisor, Otto Hutter, later Regius Professor of Physiology at Glasgow University. I was then lucky enough to obtain a post at Oxford University where I could not only set up my own laboratory but also interact with some of the best professional philosophers and mathematicians in the world.

The result is what I call a journey towards enlightenment. I use that term very deliberately, knowing that it could easily be misunderstood. There are Buddhist overtones, even a degree of presumption. How does anyone know that they are enlightened? Notice though that the word is 'towards', not 'to'. I can say that, at least, for two reasons. The first is that I know that I have moved a long way from my starting point towards a fundamental change in how I think about biological science in general. Second, I call the state I have reached enlightened because that is precisely how many others have described it, and I would certainly reject the molecular reductionist view of biology that I subscribed to in 1960. I therefore began as a reductionist biologist and have developed progressively into the very opposite: a systems biologist. That is not to deny the great value of reductionist science; but it cannot be sufficient, and this is particularly true of understanding living systems.

I now lecture frequently all over the world on systems biology. These lectures have been given to an astonishing variety of audiences, not all of them scientific. Many have been in the humanities, the social sciences, the performing arts and even religious communities, including a conference of 1,000 monks in Thailand! The reaction has been uniformly exciting, as though a cloud has been lifted. Of course, I would think that myself. Who would not think of themselves as more enlightened as they grow older? Perhaps, the first point at which I started to accept the description, at least as an approach towards a goal, was after a debate in Paris in 2008, shortly after the publication of the French translation, *La Musique de la Vie*, when I lectured to a congress of Lacanian psychoanalysts. A philosopher at that congress, Clotilde Leguil, wrote a remarkably perceptive critique both of the book and my lecture. It is on the French translation page of my website[3] and can be directly accessed on the internet.[4] That was the beginning of understanding how much my approach to the systems views of biology resonates with many other disciplines and across multiple cultures.

Buddhist overtones? Those are welcome and even deliberate. As I explain in Chapter 9, I am not religious in the usual Western sense, although many of the ideas I have expressed in my recent work seem to have resonated with religious thinkers, even to the extent of drawing me into a debate in St Paul's Cathedral in London and twice with HH The Dalai Lama. Anyway, I don't think of the tradition I feel closest to, which is the Buddhist tradition, as being, itself, a religion, at least not in the Western sense of being about a Creator of the Universe. The Heart and Diamond Sutras, for example, do not even attempt to relate a story of the creation of the universe. They form rather a philosophy of man and the limits of human knowledge. The acknowledgement of these Buddhist ideas simply came naturally as I was writing the last two chapters of *The Music of Life,* where I use a Buddhist parable to explain the concept of no-self in relation to consciousness and the nervous system, and explain how easily Western people initially misunderstood what Eastern philosophies are about. Those stories flow directly from the relativistic interpretation that I give to systems biology. Readers who wish to know more about that aspect of my thinking will find the relevant material in Chapter 9 of this book, where I explain how *The Music of Life* came to be written, and more fully in my later 2016 book *Dance to the Tune of Life: Biological Relativity,* where I explain why a systems view of living organisms is a form of the general principle of relativity of causation. Nothing exists entirely in and of its own. Causation is always relational. The biggest mistake of reductionist biology is to have ignored this basic understanding of the world.

Although this book is written as a monograph, it is important to note that I am not attempting to write a complete introduction to the mathematical modelling of the heart. That has now become a major international project, the Cardiac Physiome Project, with its own annual conferences. This book is focussed on my work, initially at UCL and then at Oxford, since I was the founder of the discipline and many of the controversies and their resolutions occurred in my own laboratory and its many international collaborators.

---

[3] https://www.denisnoble.com/french/.
[4] https://denisnoble.com/wp-content/uploads/2019/11/Clotilde-Leguil.pdf.

Acknowledgements

An earlier version of this book was published as the introductory chapters to a book published in 2012, *The Selected Papers of Denis Noble CBE FRS*. In 2024, I agreed with the publisher to edit those introductory chapters as a separate book, and I have added a chapter (10) to bring the story up to date. Anyway, today, people can readily access the original papers online. I am deeply grateful to Laurent Chaminade and Gabriel Rawlinson at World Scientific Publishing for this opportunity.

I thank Reine Bourret for extensive criticism and discussion on the draft of this book and George Ellis for his comments on Chapter 10 in relation to theories of relativity.

Recognition

Several national and international prizes have been awarded for this work on the pacemakers of the heart:

1985: British Heart Foundation Gold Medal
1991: Pierre Rijlant Prize of the Belgian Royal Academy of Medicine
1993: Baly Medal of the Royal College of Physicians
2004: Pavlov Medal of the Russian Academy of Sciences
2005: Mackenzie Medal of the British Cardiovascular Society
2008: Medal of Merit of the International Society for Heart Research
2022: Lomonosov Grand Gold Medal of the Russian Academy of Sciences

Chapter 1

Discovery of Potassium Channels and the First Heart Cell Model

1.1. Introduction

University College London (UCL) was established in the Bloomsbury area of London in 1826. The impressive central campus boasts a great classical portico resembling that of the National Gallery. That is not surprising, since it was built by the same architect, William Wilkins. The campus is surrounded by elegant Georgian squares, each with gardens in the centre. The town houses around each square are all built to the same successful design, with three main floors where the spacious living rooms of the original occupants would have been. Above these, there are small attic rooms. Below them is a basement, half underground, usually reached by a separate small set of steps to the side of the main entrance. A century earlier, these would have been used by the servants. By the 1950s, when I studied at UCL, most of these houses had been converted to other uses. Some had become residential halls for students; others were used by university departments overflowing from the crowded central campus.

In the early 1960s, as a graduate student and then a young lecturer, my life revolved around two of these Georgian houses. In one of them, I had a flat in the attic. I had become the vice-Warden of a student residence, Connaught Hall. In the other, I helped wear down the already worn stone steps to the basement. That basement housed a newly created 20th century god. It was heavily guarded by its own set of high priests. And, just as prayers are submitted on paper messages in many temples around the world, it was fed with supplications, often very long ones, written on

1

paper and rolled up like a thick scroll. Only the high priests and a select band of supplicants could read these messages. After some time, often hours of thought, the god would reply with more paper messages. Einstein would have been pleased to know that the 'mind of god' spoke in the language of mathematics.

1.2. The Mercury Computer

This god was an early electronic valve computer, programmed in gibberish code punched out in holes on rolls of paper tape. Mercury (Fig. 1.1) was built by the computer company Ferranti to a design by Manchester University, and they relied on the fact that a valve (a vacuum tube) is like a switch, with just two positions, on and off. By building banks of these switches, numbers could be represented. The racks contained around 2,000 valves and a similar number of crystal diodes. An important feature of Mercury was that it could represent floating point numbers, important for most scientific applications.

When these machines were first built in the 1940s, they were extremely expensive and very rare. So rare that the then Chairman of IBM, Thomas Watson, estimated the world market to be around five! Ferranti did a little better than that; just 19 were built. The first of these was built in 1957, just before I graduated and became a research student. They not only ate paper tape but they were also power hungry. They had five racks of power supplies. The 'priests' were, of course, a new generation of mathematicians and engineers called computer scientists.

Now that the world market in computers runs to billions, including mobile phones with processors vastly more powerful, it is hard to imagine the days when a computer was such a rarity and so lacking in function. There was no screen, no graphics, no windows, very little memory, and not even fortran coding to help the user – the programming was done in a mixture of machine code and a primitive structure called autocode. The keyboards that we used were those of teleprinters that converted type into patterns of holes on rolls of paper tape (Fig. 1.2) to be fed to the computer via a tape reader. The machines were, in effect, dedicated calculators, used only for heavy numerical work. It took hours to produce a result that, today, can be achieved in the blink of an eye on a simple laptop computer. The speed was around 10^4 flops[1] per second. Today, we can reach more

---

[1] Flop = floating point operation. The number of flops per second is often used to measure computer performance.

Figure 1.1. The Ferranti Mercury computer, *circa* 1960.

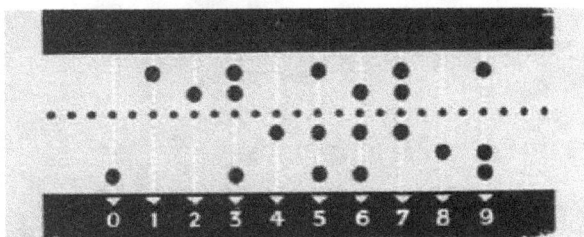

Figure 1.2. The paper tape code used by Mercury. Each row of up to five punched holes corresponded to a number.

than 10,000 trillion on the fastest computers.[2] Ten thousand billion operations can now be done in the time it took to do just one.

Time on the computer was therefore very precious. There was severe competition between the numerical analysts, the crystallographers and particle physicists, all beseeching the guardians of the machine to grant them time. The primitive coding and functionality also created a serious

---

[2] https://en.wikipedia.org/wiki/Floating_point_operations_per_second.

barrier to entry. A user not only had to understand the mathematics he was using or developing but he also had to master the difficult code. User-friendliness was an unknown quality in those days.

I believe I was the only biologist in the whole university to have dared to ask for time on Mercury. Yet, the idea that mathematics could be used in biology was an old one. William Harvey used careful calculations in his demonstration of the circulation of the blood (see Auffray and Noble, 2009). Claude Bernard in 1865 even insisted that 'the application of mathematics to natural phenomena is the aim of all science, because the expression of the laws of phenomena should always be mathematical' (see Noble, 2008). But exceedingly few biologists had followed his vision in the 100 years since his book was published. Of course, my model was the Hodgkin–Huxley nerve model published in 1952, which I had studied as an undergraduate. But that work was done with a hand calculator!

My credentials to do so looked hopeless. I had stopped studying mathematics at the age of 16, when the English school system forced an absurd choice between maths and biology. Not only did I not have a degree in maths, I didn't even have the school qualification, the A-level, that would have taught me differential equations and the maths of integration. Little wonder therefore that my initial request was declined. They clearly feared that I would be wasting scarce computer time. Yet I knew I needed to use the machine.

1.3. Experimental Discoveries

I will explain why.

It started with a major experimental discovery, first described in a *Nature* paper (Hutter and Noble, 1960) with my supervisor, Otto Hutter (Fig. 1.3). Following the lead of Hodgkin and Huxley's (1952) work on the nerve impulse, we decided to look in the heart muscle for the potassium channel current responsible for repolarisation at the end of each electrical impulse. As an undergraduate student in 1958, I had been enormously impressed with the Hodgkin–Huxley equations. Here, at last, was physiological analysis that could rival the use of mathematics in physics: precise experimental measurements of the relevant nerve electrical parameters, accurate fitting of mechanistic equations to the data, followed by integration of those equations to produce a complete explanation of the

Figure 1.3. Denis Noble and Otto Hutter at the IUPS congress in Glasgow, 1993, on the ship used for the congress boat trip up the clyde.

nerve impulse and its conduction. I was far from alone in being impressed. Just five years later, in 1963, they won a Nobel Prize for their achievement.

I naturally wondered how their work could be applied to the heart. There was a big puzzle to be solved. In the case of nerves, the permeability of the membrane[3] increases during the whole duration of the impulse. This is what we would expect, since it is generated by the opening, first of sodium ion channels and then of potassium ion channels. But, in heart cells, the reverse happens. After an initial increase in permeability, the permeability rapidly falls. Even worse, in the experiments of Weidmann (1951), who measured membrane resistance (the inverse of permeability) throughout the action potential, the permeability apparently continues to

---

[3] 'Permeability' is a measure of the speed with which substances (such as ions) cross the cell membrane. When these are electrically charged, the movement creates an electric current. The permeability can then be measured as an electrical conductance. The inverse measurement is an electrical resistance. 'Polarisation' refers to the electrical potential across the cell membrane created by these movements of ions. At rest, the inside of the cell is negative. 'Depolarisation' refers to the change towards a positive potential. 'Repolarisation' is the recovery of the resting negative potential. Many of the technical terms used in this book are explained in the Glossary.

decrease throughout the long, slow repolarisation. How could this be if it required the opening of potassium channels to bring that repolarisation about?

The experiments with Otto Hutter explained the low permeability. We measured the electrical current flowing through the cell membranes in conditions where the great majority of the current is carried by potassium ions. The result was a surprise. Contrary to the situation in nerve, the potassium ion channels that are open in the resting state immediately close when the membrane is depolarised. The graphical convention for current–voltage relations was different in 1960, so in Fig. 1.4, I have replotted those experiments using the modern convention of making the voltage the abscissa and the current the ordinate. Our results were interpreted to show the presence of two types of potassium channels. The first channel, naturally called i_{K1}, closes when the membrane is depolarised. In fact, its permeability becomes almost zero. This would explain the low permeability during the long cardiac action potential. But how, then, does

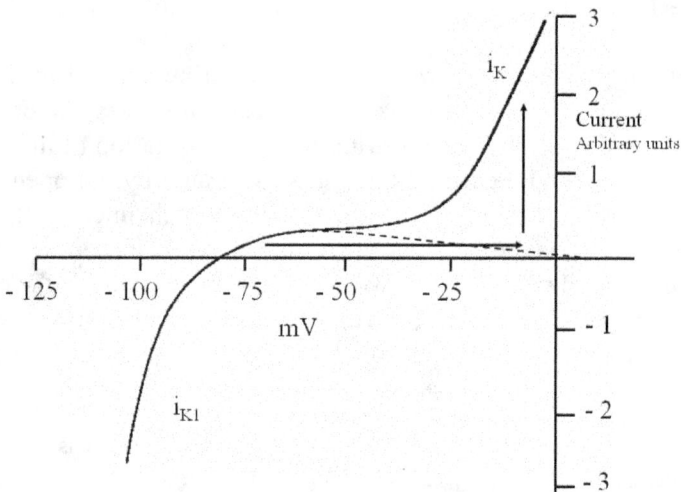

Figure 1.4. Re-drawing of the Hutter–Noble (1960) results using the modern convention that the voltage is displayed on the abscissa. The curve has also been corrected numerically for the cable properties of the fibres on which the results were obtained. The horizontal arrow indicates the first effect of membrane depolarisation, which is to rapidly reduce the potassium permeability towards zero (extrapolated dotted line). The vertical arrow indicates the slow onset of i_K.

repolarisation occur? Our results also explained that, since the second kind of potassium channel (we called it i_{K2}, but it is now called i_K) slowly increases with time, eventually overcoming the depolarising currents.

1.4. Building the Computer Model

So far, so good. These potassium channel characteristics could explain a lot of the properties of the cardiac impulse. But how can they possibly explain the slow decrease in permeability measured in Weidmann's experiments? I could see the glimmer of an explanation in the properties of the sodium channel in Hodgkin and Huxley's work. The sodium channels have two kinds of gates, one that opens the channel and another that closes it. There is a range of membrane potentials over which a small fraction of both gates are open, even after a long period of time. Could the potassium channel behaviour be taking the membrane potential slowly into this region and so producing an apparent increase in membrane resistance when, in fact, the net membrane permeability is increasing? This is the kind of counterintuitive explanation of a phenomenon that cries out for a calculation to demonstrate that it works quantitatively.

Frustrated by the initial refusal to allocate me time on Mercury, I started calculating by hand using a Brunsviga calculator (Fig. 1.5) that I found in

Figure 1.5. Brunsviga hand calculator, model 20. Numbers were entered by moving levers, and an arithmetical operation was implemented by turning the main handle. This is the machine on which Andrew Huxley performed the computations of the Hodgkin–Huxley 1952 paper. It took six months. He developed a very strong right hand. I think that is why he once beat me easily in a game of table tennis.

the physiology laboratory. I got as far as computing the rapid upstroke of the action potential (Noble, 1962a, Fig. 7). But then I did quite a different calculation. How long would it take me to go all the way through the cardiac action potential using a hand calculator? In 1952, it took Andrew Huxley 8 hours on this machine to compute 8 msec of a nerve impulse. The cardiac action potential lasts about 500 msec. So, it would require several months to do a single calculation. Clearly, although working on the hand machine taught me a lot about the process of integration (a fact that subsequently stood me in good stead when it came to programming my own fast integration routines), it was completely impractical for my project.

But, ringing in my ears were the biting comments of the guardians of Mercury: 'You don't know enough mathematics and you don't even know how to program!' In their position, I would have made exactly the same judgement. I did the only thing possible, which was to sign myself up for a course that the mathematicians were giving to the engineering students, and I bought a book (I still have it) on how to program Mercury. The maths lectures were nearly a disaster. The lecturer, Dr Few, was excellent, and he kindly agreed to mark my assignments. But I was so far behind in mathematical knowledge that I couldn't follow anything he said during the first three or four lectures. Matrices, Bessel functions and much else – they were just slavishly copied down as I panicked and wondered whether to give up. What came to my rescue was an innate talent. Although I had been required to abandon maths at the age of 16, I had not done that from choice, and up till then, I was nearly always top of the class. I must have had some natural gift for it. Slowly, just like relearning a language after rusty years of non-use, those instincts came back. I still have the assignments marked by Dr Few and his astonished comments that, after such a faltering start, I had moved to getting full marks.

Armed with this knowledge, I then returned to the guardians of Mercury, carrying my book of programming and some examples that I hoped would convince them that I was ready.

They were more sympathetic this time, so the discussion moved on to what I actually wanted to do. What, they asked, was my numerical problem, and how did I propose to solve it? Naively, I sketched out on a piece of paper the cyclical variations in electrical potential recorded experimentally, showed them the equations I had fitted to my experimental results and said that I was hoping that they would generate what I had recorded experimentally. A single question stopped me in my tracks. 'Where, Mr Noble, is the oscillator in your equations?'

Here, a word of explanation is required on the vexed question of scientific strategy. Twentieth-century biological science was largely ruled by a naïve reductionist strategy. I know because I was part of it. All we needed to do was to characterise all the components of the system we were studying. The rest would follow since, after all, it is 'just a bunch of molecules anyway'. That was my mindset too at the time. My reply should have been, 'The oscillator is an inherent property of the system, not of any of the individual components, so it doesn't make any sense for the equations to include explicitly the oscillation it is seeking to explain. That would be an empty hypothesis, not even open to the critical criterion of scientific sense, the ability to be falsified'. Instead, I continued the sketch on the paper with some hand waving that tried to indicate the cycle of activity that I thought could happen. It was all just qualitative speculation. At that time, I did not have the mathematical language for attractors, nor for multi-dimensional representations of complex systems. Even so, they were convinced to give me the two hours per day that they thought would be necessary for my project to be developed. Did I say 'per day'? I should have said 'per night' for I was offered the worst time slot: 2–4 am. The mathematically skilled astronomers, theoretical chemists and computational scientists had already grabbed more convenient times on the machine.

I now think that was the best time they could have offered me. I could work the night on Mercury and then go to the slaughterhouse at 5 am to gather the fresh sheep hearts that would form the basis of the experiments during the day. Often enough, those experiments did not finish until midnight, by which time it was necessary to have a cup of coffee and write the next modifications of the paper tapes to be fed into Mercury by 2 am. That cycle would continue for 2 or 3 days before I crashed out in a long sleep catch-up. As I explain in *The Music of Life* (2006, Chapter 6), that experience deeply disturbed my circadian rhythm, which can wander almost freely away from the norm, with the convenient result that I can travel, east or west, with minimal jet lag!

The absence of a screen, graphics, windows and all the features that make modern computers so user-friendly forced me to consider how to monitor the progress of my computations. To wait for two hours just to find that a simple error or a bad choice of a parameter had made the result useless was clearly a waste of time. But I found that this god could speak! Mercury had a loudspeaker and there was an instruction to send a pulse to the loudspeaker. It didn't take long to realise that by putting this

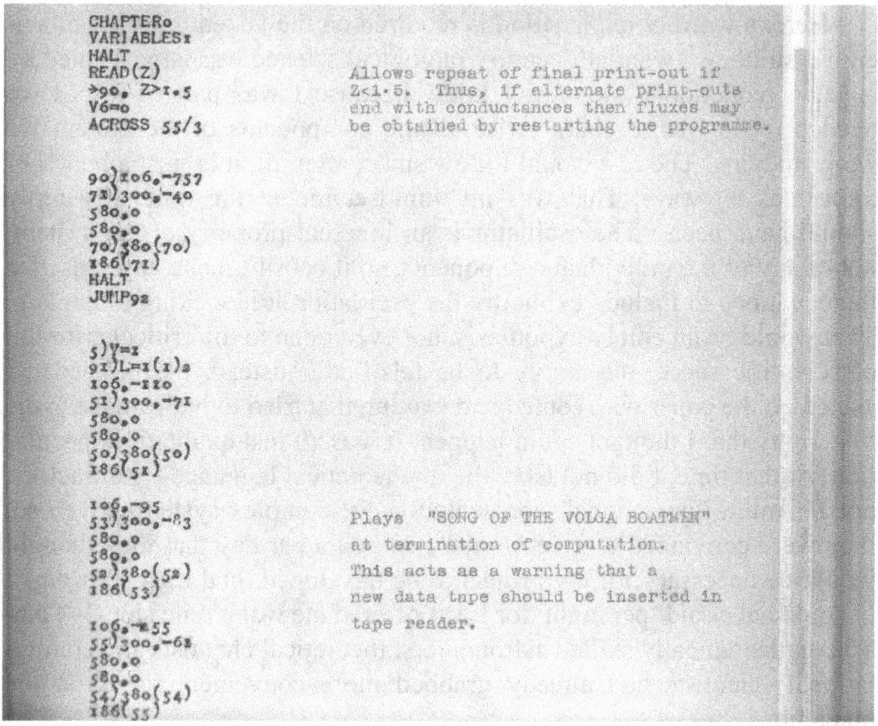

```
CHAPTER0
VARIABLES1
HALT
READ(Z)              Allows repeat of final print-out if
→90, Z>1.5           Z<1.5. Thus, if alternate print-outs
V6=0                 end with conductances then fluxes may
ACROSS 55/3          be obtained by restarting the programme.


90)106.-757
71)300.-40
580.0
580.0
70)380(70)
186(71)
HALT
JUMP92

5)V=1
91)L=1(1)2
106.-110
51)300.-71
580.0
580.0
50)380(50)
186(51)

106.-95
53)300.-83        Plays "SONG OF THE VOLGA BOATMEN"
580.0             at termination of computation.
580.0             This acts as a warning that a
52)380(52)        new data tape should be inserted in
186(53)           tape reader.

106.-*55
55)300.-61
580.0
580.0
54)380(54)
186(55)
```

Figure 1.6. Section of code for Mercury playing a melody. The lines, playing the *Song of the Volga Boatmen*, were executed at the termination of each computation and so gave a warning to feed more data to the computer (from Denis Noble's PhD thesis, 1961).

instruction into a repetitive loop, it was possible to generate notes of any duration or pitch. So, I introduced various bells and whistles to inform me of the progress of the computation (Fig. 1.6). I also wrote an error routine that played the melody 'Oh dear, what can the matter be?' I was initially told by the computer engineers that this was wasting precious computer time until I explained that this was precisely what enabled me to *save* computer time. So, I eventually made my god hum and whistle its way through the night. Once the paper tape had been fed to the computer, I could turn to my maths lectures and assignments confident that Mercury would soon tell me if something had not worked. When all was well, there would be a warbling of hunting sounds and chants, breaking out into the slow *Song of the Volga Boatmen* to warn me to be ready with the next tape of data.

1.5. The Model Works

The crazy schedule, including the maths at night and the experiments during the day, worked. A few months later, the two *Nature* publications appeared, to be followed two years later by the full paper in the *Journal of Physiology*. Rhythm did indeed emerge as an integrative property of the interactions of the components of the system, a fundamental goal of any form of system biology (Fig. 1.7).

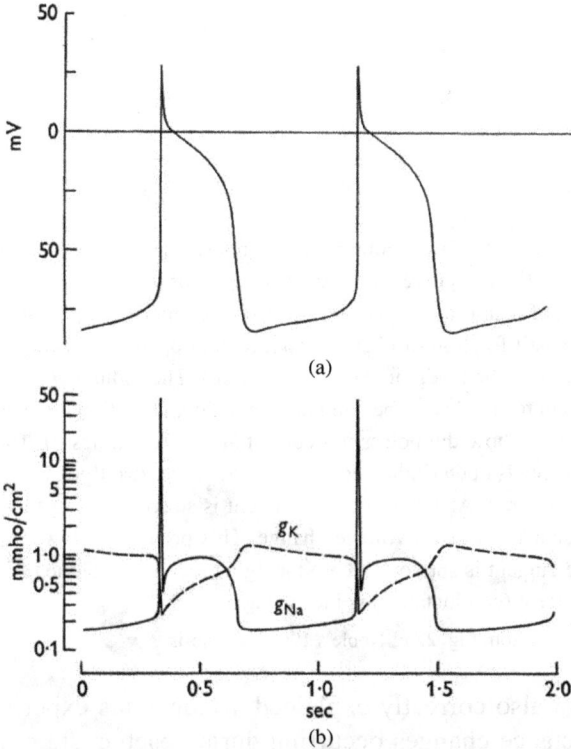

Figure 1.7. Electrical potential changes (a) and sodium and potassium conductance changes (b) computed from the first biophysically detailed model of cardiac cells. Two cycles of activity are shown. The conductances are plotted on a logarithmic scale to accommodate the large changes in sodium conductance. Note the persistent level of sodium conductance during the plateau of the action potential, which is about 2% of the peak conductance. Note also the rapid fall in potassium conductance at the beginning of the action potential. This is attributable to the properties of the inward rectifier i_{K1}, and it helps to maintain the long duration of the action potential and in energy conservation by greatly reducing the ionic exchanges involved.

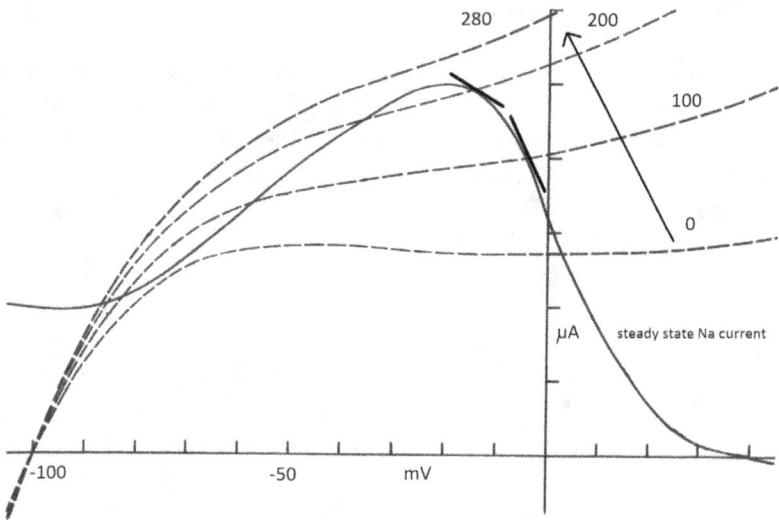

Figure 1.8.   Mechanism of the counterintuitive resistance changes during repolarisation in the heart. The continuous curve shows the behaviour of the sodium current in the steady state as a function of membrane potential. Over a wide range of potentials there is a hump attributable to a small fraction of channels that remain open. This is a general and well-understood property of the Hodgkin–Huxley equations. The sodium current is plotted here as a positive current to enable it to be compared to the opposite-flowing potassium current. The interrupted lines show the potassium current at different times (0, 100, 200 and 280 msec) as the i_K channels open (indicated by the arrow). The two thick lines are tangents to the sodium current curve. At 100 msec, the tangent is steep, meaning that a large current change would occur for a given voltage change. This produces a low 'slope resistance'. At 200 msec, the tangent is shallow, so producing a much larger slope resistance. Yet, the net membrane current (conductance) is increasing.

Source: Figure adapted from Fig. 22 of Noble's 1961 PhD thesis.

The model also correctly explained Weidmann's experimental result on the conductance changes occurring during each cycle of activity (see Noble, 1962a, Fig. 10). The delayed K current, i_K, does take the membrane potential into a region where the sodium current system displays a rapidly changing and even a negative slope resistance (Noble, 1962a, Fig. 9), so giving the impression of a decrease in permeability when, in fact, the permeability is increasing. The mechanism is revealed in Fig. 1.8.[4]

---

[4]Some general readers might wish to skip this figure and its explanation. It is sufficient to know that the experimental results could be explained by the theory (Fig. 1.9).

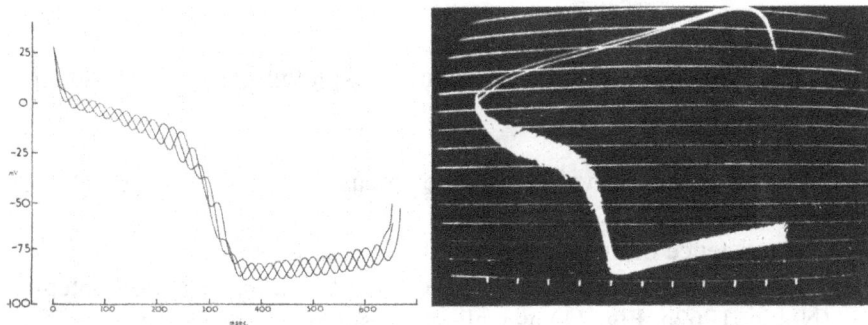

Figure 1.9. Left: Computed result using successive sinusoidal current changes to reconstruct the increase in apparent resistance during the slow phase of repolarisation, shown as a progressive increase in the width of the deflections. Right: The experimental result obtained by Silvio Weidmann using square current pulses. These illustrations were taken directly as scans from Noble's 1961 PhD thesis. The experimental result was displayed on an oscilloscope, which explains why the traces are white on a black background – this is a photo of the oscilloscope screen. The early part of the recording was also displayed as a rapid return recording on the oscilloscope. The oscilloscope photo has been scaled to use almost the same scales as the computed result.

In fact, the reconstruction was remarkably good quantitatively (Fig. 1.9). The computed and experimental results could almost be super-imposed on each other. The only difference was that Weidmann used square current pulses. For my computations, it was computationally much easier to program a sinusoidal current.

The prediction that there is a range of membrane potentials over which a 'hump' or 'window' of steady state sodium current should flow was eventually demonstrated experimentally in the Oxford laboratory (Attwell *et al.*, 1979).

1.6. Conclusions

In each of these chapters, I will draw conclusions with a balance sheet of pluses and minuses as each stage of the story develops. That will also enable the reader to see how my interpretations of systems biology have developed. What can we say about the 1962 model and its experimental basis?

1.6.1. *Pluses*

The model was surprisingly successful in explaining many experimental facts:

- The counterintuitive resistance measurements I have already referred to.
- The existence of an oscillator in the interactions.
- The existence of thresholds for the termination of the action potential (Noble, 1962a, Fig. 11) as well as thresholds for its initiation.
- The classification of the potassium current channels into two types, i_{K1} and i_K. This classification remains correct today and I will elaborate on it and how it has developed in a later chapter.
- The paradoxical effect of potassium on action potential duration (Noble, 1965), which has turned out to be important in explaining some of the changes during ischemia (reduced blood flow).

The existence of thresholds for repolarisation (item 3) was the focus of an argument that raged soon after I published the *Nature* article in 1960.[5] Johnson and Tille (1961), working in Australia, had performed an experiment that seemed to be a knock-down disproof of its existence in the ventricle, even though Weidmann (1951; 1956) had, equally unambiguously, shown its existence in Purkinje fibres. It looked as though the model was going to be disproved almost as soon as it was published. Not just disproved. The existence of these thresholds was critical to the application of any form of Hodgkin–Huxley equations to cardiac muscle. Those equations involve highly nonlinear feedback between the electrical potential and the gating of the ion channels (Fig. 1.10). To show that such feedback does not exist would clearly disprove applications of equations of this kind.

Ted Johnson was fully aware of that deep implication. If they were correct, no amount of fine-tuning of the equations would avoid this inconvenient fact. Their paper was there in *Nature* just months after my own. To say that I was worried would be an understatement. I walked around the UCL quadrangle, and backwards and forwards, to my Georgian terrace flat in a state of alarm. The ball was firmly in my court. It was either a bomb or a godsend. Which?

---

[5] General readers might wish to skip these paragraphs and jump to 1.6.2.

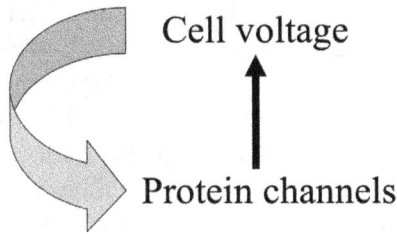

Figure 1.10.   The Hodgkin cycle (see text for description).

What was their experiment? It was clever. They had made a double-barrelled microelectrode[6] that enabled them to inject current into the same cell from which the other barrel was recording. They plunged this apparatus into the ventricle and found that, no matter how far they repolarised the membrane during the action potential, the potential simply returned to its original time course once the current pulse was terminated. Not only was there no threshold but *there was no effect whatsoever*. This is impossible to reconcile with a mechanism dependent on voltage-dependent ion channels. The title of their paper says exactly that.

Somewhere during those anxious walkabouts, I must have approached the building where Dr Few's lectures were held. Ah! Bessel functions![7] I went back to my room to find the slavish scribbles I had made when I didn't understand what he was talking about. There it was in my handwriting. If you perturb a two- or three-dimensional network, the disturbance does not decay exponentially from the source; it decays *much* faster. Engineers building cylindrical or spherical structures are familiar with the maths involved. Could it decay so fast that the effect would be nullified? I coded it all up, ran to my appointment with Mercury and tested the idea. The output was shocking. All I got was unwanted oscillations!

By that time, Andrew Huxley had moved from Cambridge to take the chair of physiology at UCL. There is a widespread impression that, because I used their equations, I must have been trained by one of them. This is not true. Alan Hodgkin examined my thesis in 1961, but I had no interaction with either of them before that. Huxley's reputation in maths

---

[6]A glass micropipette with a tip diameter less than 1 μm, but with two such pipettes stuck together so that they can be inserted into the same cell.

[7]Reminder: Bessel functions and other technical terms will be found in the Glossary.

was phenomenal, so I went to him with my oscillatory output from Mercury and asked for his advice. It was swift. 'You need Bessel functions of an imaginary argument'. *Imaginary* argument? I had no idea what an imaginary argument was; I couldn't even imagine what it *might* be. Yet, it is fundamental to the very nature of mathematics that it does not have to deal only with what we imagine to be reality. Many important mathematical developments use constructs that we find difficult to imagine as objects in the real world. It is easy enough to imagine the numbers from 0 to ∞ arranged on an infinite line. We can even extend this line backwards to −∞ to include negative numbers. But suppose we expand the line of numbers to become a two-dimensional sheet. This can then include a line at 90° to that of our 'real' numbers. To what could this possibly correspond? The answer is fascinating. It can be represented as being the set of 'real' numbers multiplied by an 'imaginary' number, which is $\sqrt{-1}$. The sense of 'imaginary' is now obvious. Any real negative number multiplied by itself would always produce a positive number, yet $\sqrt{-1}$ needs to be a number which, if multiplied by itself, would produce a negative number. Not only is this fascinating mathematically but it turns out to produce mathematics that has important practical applications. This is true for Bessel functions.

So, it was back once again to the notes of Dr Few's lectures and a quick climb up the learning curve on $\sqrt{-1}$ and all its ramifications. Then back to coding Mercury. That night, I must have emerged from the basement of God singing its praises. Using Bessel functions of an imaginary argument for the spread of current flow from the source, and my equations for the membrane channels, the result was exactly what Johnson and Tille had found. I wrote the paper in a frenzy. It was published in the newly established *Biophysical Journal* (Noble, 1962b) and it showed unambiguously that the Bessel function decay is sufficiently rapid to completely linearise the current–voltage relations even in a highly nonlinear system (see Fig. 3 in that paper and Fig. 3.3 in Chapter 3 of this book).

That experience was not only absolutely critical to the development of the theory of cardiac electrophysiology but it also introduced me to the theory of electric current flow in excitable cells in general. That became the title of a major book (see Chapter 3). Meanwhile, the 1962 model had survived the most fundamental challenge to its basis: fundamental, because that challenge could not have been met simply by further developments of the model. It required a mathematical proof that the result was a function of the way in which the current had been applied.

Perhaps the 1962 model was too successful. It found its way rapidly into an edition of Hugh Davson's famous *Textbook of General Physiology* far more quickly than for its own good. But by the time that edition of his book was published in 1970, the cracks were already appearing. I was somewhat shame-faced when I saw it all in Davson's book. He never asked me about it. I could have told him that calcium channels had already been discovered (by Harald Reuter) and that the model was ready for a major update at least. I will describe that update in Chapter 2. Textbooks have a hard time keeping pace with a rapidly developing field! This one was developing at breakneck speed.

1.6.2. *Minuses*

Those cracks are large:

- The model made the sodium current equations perform all the roles that were soon to be revealed to also be attributable to calcium channels (Reuter, 1967). In retrospect, this failure can also be seen to anticipate the findings that sodium channels are in fact different in nerve and cardiac cells.
- There was no representation of changes in ionic concentrations. These subsequently turned out to be critical to one of the next stages of development (see Chapter 5).
- There was nothing in the model corresponding to the time-dependent pacemaker channels to be discovered later, first under the guise of a new form of i_{K2}, then later as the i_f channel. So, the pacemaker rhythm in Purkinje tissue (for that was the part of the heart I was working on) was attributed to the wrong mechanism. It was, however, one of the mechanisms that eventually turned out to be correct for the real pacemaker tissue of the heart.

In relation to the failures, it is important to note that, at each stage in this story, we will find that the failures often revealed more than the successes. It should be remembered that it is the function of a scientific hypothesis to be falsifiable. If the model was developed well and on the basis of the best interpretation possible at the time, we should be celebrating the failures at least as much as the successes. To anticipate a later discussion in this book, there was no circularity in the thinking. Each model was a stage in the continuous iteration between theory and experiment which is also a fundamental characteristic of systems biology.

1.6.3. *Contributions to systems biology*

In each of these chapters, I will also highlight the insights that were gained from a systems approach to biology.

1.6.3.1. Downward causation

As Alan Hodgkin had already pointed out in the case of the nerve, electrical excitation involving voltage-dependent ion channels can be seen as an example of a cycle (originally called the Hodgkin cycle) in which ion channels contribute charge to determine the cell potential, which in turn influences the channel kinetics themselves. It is characteristic of such cycles that their behaviour requires quantitative analysis since intuition often fails to see the important properties of the whole system.

The feedback from the cell voltage to the ion channels is an example of what, in *The Music of Life*, I call 'downward causation' since it involves an influence of higher levels (in this case the cell as a whole) on lower-level processes, in this case the ion channels.

1.6.3.2. Energy conservation

The main integrative insight that the i_{K1} discovery led to was an energy-saving mechanism. The flow of ions down their electrochemical gradients must eventually be reversed by processes that consume energy. A long action potential can be generated without the i_{K1} mechanism, as shown by Fitzhugh (1960) working on the Hodgkin–Huxley equations at the same time as I was working on my first models. But the price is substantial in terms of energy demand. By greatly reducing the permeability during the long action potential, only relatively small ionic currents flow during most of the time, so conserving energy. It is easy to see why channels like i_{K1} were selected during the evolutionary process.

1.6.3.3. Genetic program

I can't conclude this chapter without drawing attention to the way in which the feeding of paper (or magnetic) digital tape into a computer influenced thinking about the role of the genome in biological systems. Monod and Jacob introduced their famous description of the genome as a genetic program using precisely this analogy. Jacob was quite specific about it: 'The programme is a model borrowed from electronic computers. It equates the genetic material with the magnetic tape of a computer' (Jacob, 1982). I think that this analogy is misleading. I will return to this question in Chapter 9 where I introduce my book *The Music of Life* (Noble 2006).

References

Attwell, D., Cohen, I., Eisner, D., Ohba, M. and Ojeda, C. (1979) 'The steady state TTX sensitive ("window") sodium current in cardiac Purkinje fibres', *Pflügers Archiv, European Journal of Physiology*, 379, pp. 137–142.

Auffray, C. and Noble, D. (2009) 'Conceptual and experimental origins of integrative systems biology in William Harvey's masterpiece on the movement of the heart and the blood in animals', *International Journal of Molecular Sciences*, 10, pp. 1658–1669.

Fitzhugh, R. (1960) 'Thresholds and plateaus in the Hodgkin-Huxley nerve equations', *Journal of General Physiology*, 43, pp. 867–896.

Hodgkin, A. L. and Huxley, A. F. (1952) 'A quantitative description of membrane current and its application to conduction and excitation in nerve', *Journal of Physiology*, 117, pp. 500–544.

Hutter, O. F. and Noble, D. (1960) 'Rectifying properties of heart muscle', *Nature*, 188, p. 495.

Jacob, F. (1982) *The Possible and the Actual*. New York: Pantheon Books.

Johnson, E. A. and Tille, J. (1961) 'Evidence for independence of voltage of the membrane conductance of rabbit ventricular fibres', *Nature*, 192, p. 663.

Noble, D. (1960) 'Cardiac action and pacemaker potentials based on the Hodgkin-Huxley equations', *Nature*, 188, pp. 495–497.

Noble, D. (1962a) 'A modification of the Hodgkin–Huxley equations applicable to Purkinje fibre action and pacemaker potentials', *Journal of Physiology*, 160, pp. 317–352.

Noble, D. (1962b) 'The voltage dependence of the cardiac membrane conductance', *Biophysical Journal*, 2, pp. 381–393.

Noble, D. (1965) 'Electrical properties of cardiac muscle attributable to inward-going (anomalous) rectification', *Journal of Cellular and Comparative Physiology*, 66 (Suppl 2), pp. 127–136.

Noble, D. (2006) *The Music of Life*. Oxford: Oxford University Press.

Noble, D. (2008) 'Claude Bernard, the first Systems Biologist, and the future of Physiology', *Experimental Physiology*, 93, pp. 16–26.

Reuter, H. (1967) 'The dependence of slow inward current in Purkinje fibres on the extracellular calcium concentration', *Journal of Physiology*, 192, pp. 479–492.

Weidmann, S. (1951) 'Effect of current flow on the membrane potential of cardiac muscle', *Journal of Physiology*, 115, pp. 227–236.

Weidmann, S. (1956) *Elektrophysiologie der Herzmuskelfaser*. Bern: Huber.

Chapter 2

Discovery of Multiple Slow Channels

2.1. Introduction

One of the consequences of my move to Oxford in 1963 was that I soon discovered that I did not need to spend time looking for funding for people to come to work in my laboratory. Almost before I had time to establish a laboratory, young scholars from all over the world, coming to Oxford on fully funded scholarships, were approaching me to join the laboratory. For around 12 years, I did not need to write grant requests to support a research team. Then, for a further 10 years, from 1975 until 1985, I held a long-term programme grant from the Medical Research Council that enabled me to update and extend the laboratory facilities. But still, I had people applying to join the lab with their own full funding.

Just one example is the Rhodes Scholarships, which bring brilliant graduate students to Oxford University from various parts of the world, including some of the Commonwealth countries. Many of them also come from the United States.[1] In the 1960s, the North American students made the journey on the Queen Elizabeth liner sailing across the Atlantic Ocean to be brought to Southampton before the bus journey to Oxford, where they were taken to each of their colleges. The 1966 intake included two who came to work in my already crowded laboratory: Stephen Bergman from Harvard and Dick Tsien from MIT. I already had three other research students on prestigious international scholarships.

---

[1]This is not a nostalgic throwback to when the American states were British colonies. Germany is also allocated Rhodes Scholarships.

Yet, leaving UCL in 1963 was a heart-rending decision. The department there was stacked full of Nobel Prize winners, including AV Hill, Bernard Katz and Andrew Huxley. It was a clear world leader in cell biophysics. Those were the last glory days of cell biophysics, as the focus of the Nobel Prize in physiology and medicine started to shift firmly towards the spectacular growth of molecular biology. The seismic shift towards fully gene-centric biology was well underway.

I had also spent some of my time at UCL interacting with philosophers like AJ Ayer (who moved from UCL to Oxford a year after I graduated in 1958) and Stuart Hampshire (still at UCL when I left). My intellectual wings were rapidly spreading way beyond the laboratory. I think that once people have tasted the academic joys of multi-disciplinary work, the temptation to go further afield is irresistible.

Oxford therefore had two attractions for me. First, there was the exciting challenge to create a laboratory of cell biophysics and cardiac physiology, neither of which existed at Oxford, despite the fact that the first professor of physiology in Oxford had been John Burdon-Sanderson, the discoverer of the long duration of the cardiac action potential (Burdon-Sanderson and Page 1883). After his work, heart and cell biophysics had been neglected at Oxford, which became dominated by neurophysiology (Sherrington, Liddell, Phillips) and respiration (Douglas, Lloyd, Cunningham).

The second was the phenomenal reputation of Oxford in philosophy. I had already read books and papers by AJ Ayer, Gilbert Ryle, JL Austin, Peter Strawson, Richard Hare and Bernard Williams, some of the seminal Oxford philosophers. Not long after moving to Oxford, I was already engaging some of them in debate and publications (Noble, 1966; 1967a; 1967b), including substantial interactions with Anthony Kenny, Charles Taylor and Alan Montefiore. The significance of these interactions and the grounding in professional philosophy that this gave me will become clear in Chapters 8 and 9 of this book. Those interactions played an important role in my journey towards a systems approach to biology.

2.2. The Nerve Group in the Noble Lab

But let's first return to the Rhodes and other prestigious scholars of the 1960s. Stephen Bergman (Fig. 2.1) came to work on the neurophysiology of learning and memory in insects. Dick Stein (also from MIT, as a

Figure 2.1.   Denis Noble with Stephen Bergman at Balliol College in the early 1970s before a dinner speech by the former Prime Minister Harold Macmillan. Asked by Stephen how to give a speech, Macmillan replied, 'First, remember that a speech is about just one thing. Second, start with a surprise – don't tell them what they think you will tell them at the beginning'. He reduced the Hall to tears as his speech recalled the devastation of the First World War. My father, George, was wounded in that war, in the battles of the Somme.

Marshall Scholar) had rapidly established work with me in nerve physiology, and we published substantial papers on nerve excitation (Noble and Stein, 1966) that owed a lot to my experience of the argument on excitation and repolarisation thresholds described in Chapter 1. In fact, Dick Stein also built a computer in the laboratory, designed for neurophysiological work, a phenomenal achievement in those days. Keir Pearson was also working on the control of movement. Both of them eventually took tenured posts in Canada at the University of Alberta. Stephen Bergman was the last in this line of nervous systems researchers in my lab. He then became an MD, then a novelist and wrote the best-selling novel

The House of God, on which, 30 years later, I wrote a commentary (Noble, 2008) since it has become a classic cult book and almost mandatory reading for medical students. As the author Samuel Shem, Steve created a whole new genre of fiction, focussed on medicine, hospitals and what health care should be, but isn't, doing.

2.3. Establishing the Oxford Cardiac Lab

The reason that the neural part of the laboratory I had established with Dick Stein died out after that has a lot to do with what I will call the phenomenon of Dick Tsien (Fig. 2.2). There simply wasn't room for both areas when the cardiac area was expanding to tackle some very difficult and contentious problems. Anyway, the tradition in nerve biophysics was soon taken over in Oxford by Julian Jack (see Chapter 3). The reputation that my laboratory has enjoyed in cardiac electrophysiology ever since then was firmly established by the work that Dick Tsien did during his thesis research. Out of that work, we published around 10 papers together and an enormous book (see Chapter 3) on the mathematics of excitable cells.

Figure 2.2.   Denis Noble with Dick Tsien in Montpellier at the symposium celebrating Denis's retirement in 2004.

Not surprisingly, he went on to become a Member of the National Academy after greatly extending his reputation both in cardiac and neural science following his return to the US.

The field of cardiac electrophysiology was, in any case, being revolutionised by the introduction of the voltage clamp technique. The main criticism of my experimental work leading up to the 1962 model was that it had all been done without control of the membrane potential. Since this was also what controlled the ion channels (the 'downward causation' in the Hodgkin Cycle described in Chapter 1), it was seriously limiting to control the current but not the voltage.

The first successful voltage clamp technique in the heart was achieved in Wolfgang Trautwein's laboratory in Heidelberg (Deck and Trautwein, 1964). My supervisor, Otto Hutter, was also involved in developing the method at the same time (Hecht *et al.*, 1964). These developments occurred while I was busy building the apparatus for my Oxford laboratory (we built virtually all our own electronic equipment in those days), so I was two years behind when the technique was first used in my own lab (McAllister and Noble, 1966; 1967).

By then, it was already clear that there was a central and challenging puzzle to be resolved. Despite the experimental evidence described in Chapter 1, neither the early Trautwein work (Dudel *et al.*, 1967) nor the Hecht *et al.* experiments showed the delayed potassium current, i_K! This was a serious, possibly even fatal, challenge to the 1962 model, or indeed any model dependent on such mechanisms for repolarisation. Today, the story of i_K is so completely accepted as a key process in repolarisation that it is hard now to realise how difficult it was to establish it. What then was the problem and the controversy?

Hecht *et al.*'s work can now be seen in retrospect as revealing a completely different ionic current, now called the transient outward current, i_{to}. I suspect that this channel simply masked the onset of i_K in their work in 1964. As to why the Trautwein group had difficulty in recording it initially, the best explanation I can offer is that, as the work in this chapter shows, there are many different potassium-carrying channels.[2] Dissecting them out as components of electric current was difficult, requiring patient hand-computational analysis of multi-exponential processes. And, as the

---

[2] There are now known to be over 80 mammalian genes forming templates for subunits of potassium channels. Potassium channels form the largest family of ion channel proteins. Not all of these are expressed in the heart.

McAllister and Noble (1966) paper acknowledges (summary item 4), 'in some fibres, no delayed rectification is observed even when the membrane potential is made positive' (see also the comments on page 212 of the Noble and Tsien, 1968 paper). We suspected that the process of dissecting out the short Purkinje strands may have damaged some of their functionality. The Trautwein group did eventually record the i_K current, but it was the early voltage clamp experiments with Eric McAllister (McAllister and Noble, 1966) that fully confirmed its existence. We also showed that sodium ions were critical for observing the current changes at very negative potentials. Chapter 5 will reveal the explanation of this surprising but significant finding.

The key to my own lab's research in working out how to detect slow potassium channel currents in voltage clamp experiments came from work with Eric McAllister (also a Rhodes Scholar). We argued that, if i_K was responsible for the pacemaker depolarisation in Purkinje fibres, then why not apply voltage control in the relevant voltage range, negative to about –60 mV? This had the additional advantage that we could avoid the technical difficulties of controlling the voltage once the threshold for the fast sodium current had been passed. We succeeded in recording the expected slow current changes, but we did not know then that it was a completely different current and that it was critical that we had kept sodium ions in the bathing solution!

2.4. The Phenomenal Arrival of Dick Tsien

This work with Eric McAllister paved the way for the first major contribution of Dick Tsien. The 1968 paper with him was a very ambitious one since we aimed to produce precisely the same degree of rigour in the experimental measurements of the channel kinetics as Hodgkin and Huxley had achieved in the squid axon. Yet, the cardiac fibres on which we were working were very small compared to the size of the squid giant axon (roughly 50 μm compared to up to 1000 μm). Instead of inserting a highly conducting metal wire along the inside of the fibre, which was possible for Hodgkin and Huxley, we had to access it with fine glass pipettes called micro-electrodes. These differences seriously limited the speed with which we could control the membrane potential because the required current had to pass through a very high resistance at the tip of the electrode. Nevertheless, we succeeded in analysing the slow current changes in the pacemaker range in great detail, including the kinetics (Fig. 2.3),

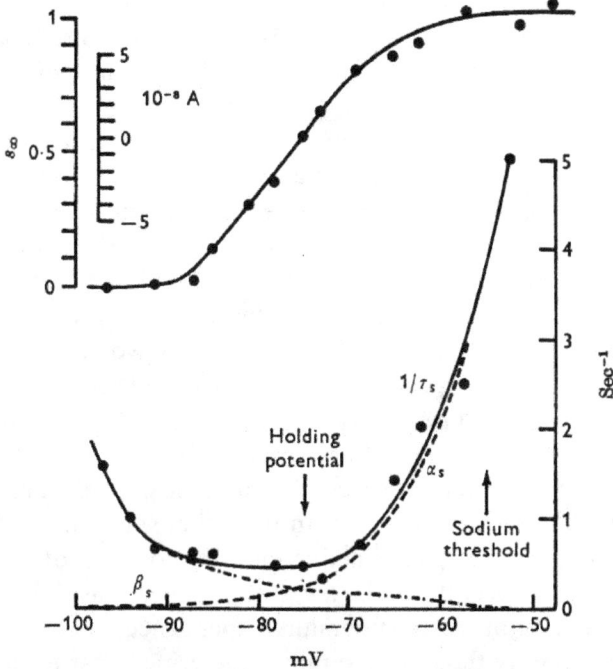

Figure 2.3. Kinetics of the slow ion channel current activated in the pacemaker range of potentials. This was the first complete analysis of the gating kinetics of a cardiac ion channel. In the 1968 paper, it was interpreted as a potassium channel activated by depolarisation. Subsequent work (see Chapter 5) showed that it is a combined sodium and potassium channel activated by hyperpolarisation. The activation curve (top figure) should therefore be the other way round.

Source: Adapted from Fig. 4 of Noble and Tsien, 1968.

voltage dependence, the transfer function – conceived as the ion flux in the absence of gating (Tsien and Noble, 1969) as shown in Fig. 8 of Noble and Tsien, 1968 – and the reversal potential. We found several completely new results.

First, the application of transition state theory to Hodgkin–Huxley-type channels in Tsien and Noble (1969) in the newly established *Journal of Membrane Biology* clarified the nonlinear behaviour of ion channels. It was already clear that the gating of ion channels produces highly nonlinear behaviour. We highlighted the possibility that an ion channel might display nonlinear behaviour even *in the absence of the main kinetic gating*. This idea later turned out to be correct. One of the major components of i_K shows strong nonlinearity similar to what I originally found for i_{K1}. This is clear in Fig. 9 of the Noble and Tsien (1969a) paper. The fast component (what we called i_{x1}, but is now called i_{Kr}) is nonlinear, while the slow component (i_{x2}, later called i_{Ks}) is linear.

The terminology here gets very confusing because it changed as the story developed, so I have included a table (Table 2.1) summarising the various components and how their names changed in subsequent work.

Second, we showed that the energies involved in the gating of the channel in the pacemaker range were unusual: instead of the temperature dependence (expressed as a Q_{10} – the change in speed for a 10°C change in temperature) falling in the expected range of 2–3, it was around 6, more than twice that expected for an ionic channel gated by voltage (see Fig. 12 of Noble and Tsien, 1968). This is one of the highest values known for an ionic channel. Expressed as an activation energy, this means that the reaction would never proceed on a biologically relevant time scale without something to help it do so. In transition state theory, that would require a negative entropy of activation. The origins of this are still unknown. I now suspect it may be a systems property, as linked reactions, each having a more modest temperature dependence.

Most importantly though, we applied the critical test for a potassium channel current. This is that the potential at which the current flow reverses should change with potassium ion concentrations, following the Nernst equation. The results passed this test apparently very well indeed

Table 2.1.   Summary of the channels carrying potassium ions discovered in the early work and their current designations. For reasons that will be elaborated in Chapter 5, what was originally called i_{K2} became two channel systems activated in completely different voltage ranges (plateau range and pacemaker range). The mechanisms in the plateau range became further subdivided. The existence of the two distinct voltage ranges was fully established in Noble and Tsien, (1969b) – see their Fig. 7.

Channel	Original name	Current name
Resting K channel	i_{K1}	i_{K1}
Delayed K channels	i_{K2} (plateau range)	i_K
Fast	i_{x1}	i_{Kr}
Slow	i_{x2}	i_{Ks}
'Pacemaker' channel	i_{K2} (pacemaker range)	i_f
Transient K channel	i_{to}	i_{to}

(pages 195–196 of Noble and Tsien, 1968). Figure 5 of that paper shows how 'clean' this reversal often appeared in the results. We had no idea at that time that nature had set a fabulous trap for us and that the Nernst equation test might be faulty.

This work led to the first reinterpretation of the mechanism of the pacemaker depolarisation. Dick Tsien did the computation shown in Fig. 13 of the 1968 paper. Together with the papers with McAllister, this clearly established that the pacemaker depolarisation was attributable largely to a channel whose voltage-dependent gating occurred *within the pacemaker range of potentials* (see Fig. 1), not at strongly depolarised levels as in the 1962 model. This was the first stage in revising that model.

Clearly then, there are two very different voltage ranges in which slow, potassium-dependent current changes occur. One lies in the voltage range (−90 mV to −60 mV) of the pacemaker depolarisation; the other in the voltage range (positive to −50 mV) at which the slow phase of repolarisation occurs. The 1968 paper took the analysis of the new mechanism in the pacemaker range as far as we could at that time. We also knew why it was not recorded in the 1960 experiments. Its sodium-dependence meant that in the sodium-deficient solutions used in the 1960 experiments, *it was completely absent*. I will return to the further analysis of this mechanism in Chapter 5.

2.5. Repolarisation Mechanisms

Having, it seemed, sorted out the situation in the pacemaker potential range, the scene was now set for tackling the channels activated in the depolarised range of potentials, i.e. the equivalent of the original i_K in the 1962 model. As can be seen by the problems encountered by the Trautwein group (Dudel *et al.*, 1967) and by Hecht *et al.* (1964) even in just recording the existence of the currents, this was not an easy project.

Before I describe the results of the 1969 paper with Dick Tsien, I should explain some technical problems. Figure 3 in that paper will serve as an example. Part of this figure is reproduced in Fig. 2.4. Each current trace lasts about 20 seconds. To achieve steady state before recording the next one, we had to wait for a minute or more (see McAllister and Noble, 1966, Fig. 9 for more information on this slow recovery time). To collect all the traces (around 60) shown in that figure – which were, in any case, laboriously pasted together from continuous pen recorder traces – required

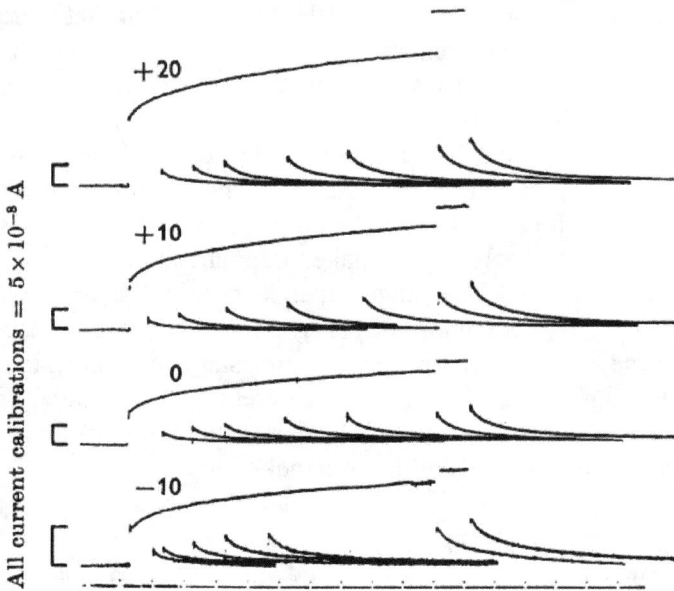

Figure 2.4.   Part of the extensive set of recordings of slow potassium channels in the plateau range of potentials. The membrane was held at −30 mV and then depolarised to the potentials indicated for varying periods of time to record both the onset and decay of ionic current as the protein channels open and close. The time scale at the bottom is seconds, so each of these recordings required many tens of seconds to occur and even longer periods of rest to allow full recovery.

Source: Adapted from Noble and Tsien, 1969a, Fig. 3.

around 4 hours.[3] Keeping two electrodes inside the heart cells for that length of time, and keeping the cells themselves in a stable state for such a long period of time, was extremely difficult. In the conditions used today with patch clamp recording from single cells (rather than multicellular tissue) and with automated computer recording, it is difficult to realise what an achievement this was. I freely acknowledge that this was attributable to Dick Tsien's skill and persistence. I could hardly believe the value of the treasure he brought to be analysed when this experiment was completed. Great rolls of pen recorder paper contained the data. Weeks were needed to analyse it because it had to be done by hand-fitting to multi-component variations. This is also a suitable point to remind

---

[3] Dick Tsien tells me, 40 years later, that he had to miss an important personal meeting to complete this extraordinary experiment.

readers that, in those days, many journals, like the *Journal of Physiology*, insisted on alphabetical listing of authors. Dick was the major author of that and the 1968 paper. In today's conventions, he would have been the first author.

The analysis was lengthy for another reason as well. Even a cursory glance at the recordings shows that the changes do not follow a single exponential time course. Figure 2.5 shows this clearly by plotting some of the data as bi-exponential components (another form of analysis that had to be carried out entirely by hand). The results of the analysis were clear; with some minor exceptions at strong and long depolarisations, all the results were consistent with the existence of two types of channel underlying i_K changes, one faster than the other. So, the two main components of i_K were identified. We called them i_{x1} and i_{x2} for reasons that need not concern us here, but they are clearly what later workers (for example, Sanguinetti and Jurkiewicz, 1990) called i_{Kr} and i_{Ks}. The work with Dick Tsien preceded those studies by more than 20 years. The only significant difference between the results is that ours were obtained on multicellular tissue, whereas later recordings used patch clamp on single cells.

Armed with this detailed analysis of the channel kinetics and other properties, we could revisit the calculations of the repolarisation phase. Dick performed the numerical integration following a procedure similar to that used in the 1968 paper for the pacemaker depolarisation. The result was impressive and is shown in Fig. 2.6. In particular, it explained why

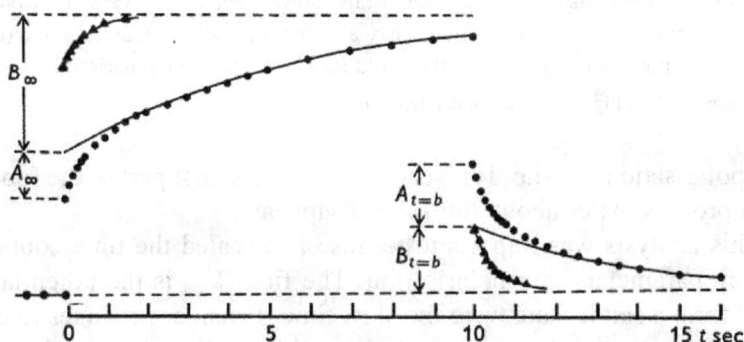

Figure 2.5. Analysis of one of the recordings of slow current changes in Fig. 2.3, showing the presence of two components (A and B), each with exponential time courses. These were referred to as x_1 (A) and x_2 (B) in Noble and Tsien (1969a) and are clearly what we now call i_{Kr} and i_{Ks}.

Source: From Noble and Tsien, 1969a, Fig. 4.

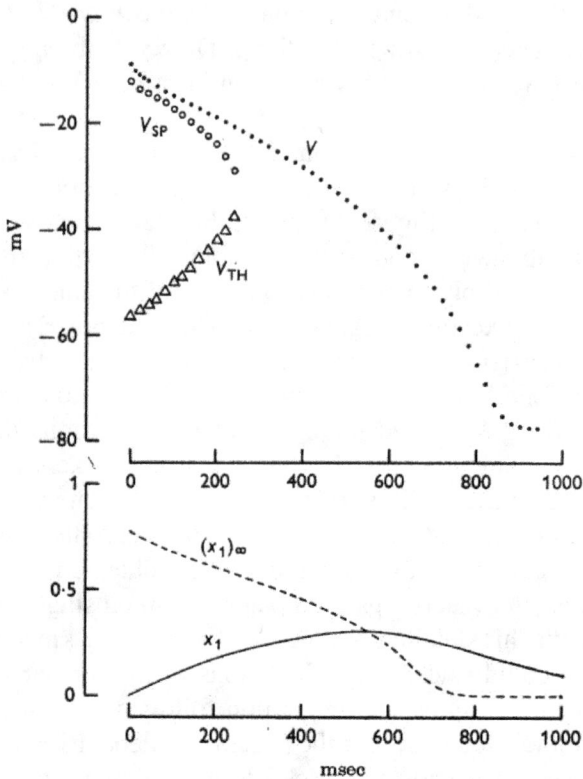

Figure 2.6.   Reconstruction of the variation in activation of i_{Kr} (here called i_{x1}) during the action potential repolarisation phase. The results showed that around 35% of the channels become activated during repolarisation. They also reproduced the changes in the (quasi-) stable plateau potential (V_{SP}) and the threshold for all-or-nothing repolarisation (V_{TH}).

Souce: Adapted from Fig. 3 of Noble and Tsien (1969b).

the repolarisation threshold exists only during the first part of the repolarisation process. After about 200 μs, it disappears.

This analysis was important because it revealed the time courses of two key parameters in repolarisation. The first, V_{SP}, is the potential that would form a stable state were the slow time-dependent changes to cease. This is the potential towards which the cell is tending. During the period for which this point exists, there is a kind of stability. We call it a quasi-stable state since it is also, itself, slowly moving. The second, V_{TH}, is the threshold point for initiating all-or-nothing repolarisation. These two points slowly approach each other and then they disappear together. After that time, about halfway through the action potential, the repolarisation process becomes less stable. The voltage time course is then best described as a 'freefall', or to use another terminology favoured by some cardiac electrophysiologists, there is no 'repolarisation reserve' left. The system can no longer resist perturbations in either direction. This is an insight that is fundamental to the understanding of cardiac arrhythmias that occur through failure of the later parts of the repolarisation process. I will return to this issue in Chapters 7 and 8.

2.6. Development of the McAllister–Noble–Tsien Model

The success in reproducing both the pacemaker depolarisation and the plateau phase of the action potential naturally led to the question whether a complete revision/replacement of the 1962 model could be attempted. But it took some time to achieve this, partly because it was necessary to incorporate equations for several other channel mechanisms that had been discovered elsewhere, such as the calcium channels first discovered by Harald Reuter and the transient outward current first observed by Deck *et al.* (1964). The McAllister–Noble–Tsien model eventually appeared in 1975 (McAllister *et al.*, 1975). As we will see in Chapter 5, it contained a major fault in attributing what was then called i_{K2} to a pure potassium channel, but the work that went into creating it was meticulous and highly accurate. As an indication of the accuracy, I include Fig. 14 from that paper as Fig. 2.7. This is a reconstruction of the beautiful experiment by Weidmann (1951), applying short current pulses during pacemaker depolarisation.

This figure was calculated by Otto Hauswirth, an Austrian pharmacologist working as a postdoctoral fellow in my laboratory. The result is extraordinarily accurate, and it is counterintuitive since one would expect depolarising pulses to *shorten* the interval to firing and hyperpolarising pulses to *lengthen* it. Exactly the reverse applies since the voltage deflections produced by the pulses directly influence the channel gating in this range of potentials. This is the kind of counterintuitive result in cardiac electrophysiology that has frequently required computation to reveal what is happening. The accuracy of the reconstruction also shows that the measurement of the gating characteristics was good. How could this be so accurate when the model itself was not correct in the identification of the ions carrying current through this channel? I will defer the answer to that question to Chapter 5.

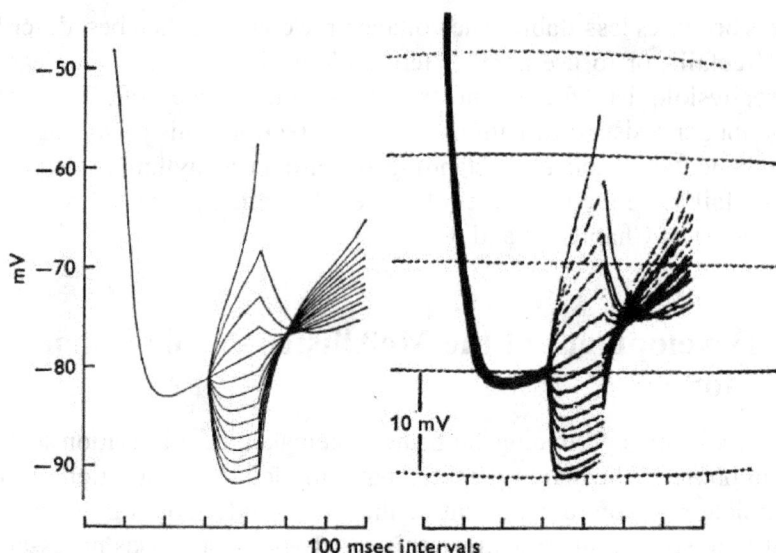

Figure 2.7. Chronotropic effects of short current pulses applied in the pacemaker range of potentials. Left: Computed effects of depolarising and hyperpolarising current pulses applied during the pacemaker potential. Note that subthreshold depolarising currents slow the subsequent approach to threshold, whereas hyperpolarising pulses speed the subsequent depolarisation (computed from the MNT model by Hauswirth). Right: Experimental records obtained by Weidmann (1951).

I will finish the general description of the articles relevant to this chapter with reference to the paper by Hauswirth, Noble and Tsien, published in 1968 in *Science*. This article shows a novel action of a hormone/transmitter. Adrenaline (epinephrine) had already been shown to activate ion channels: both i_{Ca} and i_K are up-regulated by adrenaline. What this paper shows is that it can also shift the voltage dependence of the gating mechanisms. Dick Tsien subsequently followed this discovery up by showing that the regulation occurs via intracellular cAMP (Tsien *et al.*, 1972).

2.7. Conclusions

2.7.1. *Pluses*

The discovery of multiple components of the potassium channels involved in repolarisation has major significance for work on drugs and other agents that cause arrhythmia. This work therefore opened up the later extensive collaborations with the pharmaceutical industry (Chapter 4).

The McAllister–Noble–Tsien (1975) model was a major development, which has languished in the shade largely because the tumultuous events to be described in Chapter 5 occurred only a few years after its publication. But it should be recognised for the accuracy and detail that was involved, largely attributable to Dick Tsien's work, both experimental and theoretical. I have included Fig. 5, showing the chronotropic effects of current pulses in the pacemaker range to illustrate that fact. A beautiful theory, however, only requires one ugly fact to destroy it. It was painful to see this one destroyed in that way, even though, as I acknowledge repeatedly in these accounts, we should be celebrating the failures of theories as much as the successes, particularly when those failures reveal major insights.

2.7.2. *Minuses*

Clearly, the big minus in this work is that we missed the warning signs that the channel identified in the pacemaker range of potentials was not a pure potassium ion channel. The warning signs were its sodium dependence (but recall that many channels are controlled by ions other than those they transport) and that the fit to the Nernst equation for the reversal potential was both highly accurate (the slope of the line against external potassium concentration was virtually exactly 60 mV) and always a little too negative. That story will be continued in Chapter 5.

2.7.3. *Contributions to systems biology*

An essential feature of successful systems biology involving computational models is the iteration between theory and experiment. The McAllister–Noble–Tsien model was an essential stage in that iteration and so laid the groundwork for the next stages in the cycle. In fact, one of those followed quite rapidly since the first model of ventricular muscle cells was that of Beeler and Reuter (1977), which was essentially a modification of the McAllister–Noble–Tsien model. The Beeler–Reuter model was later to be the starting point for the Luo–Rudy models of ventricular cells (Luo and Rudy, 1991, 1994a, 1994b). Many of the insights of both of these models have been carried forward into subsequent developments.

That is particularly true of the later generations of models developed in my laboratory.

This period of work laid the foundations for the classification of cardiac potassium channels. All the main types of channels were revealed. This advance has tended to be obscured by later work, partly through changes in nomenclature but also through identification of the genes involved, of which there is now a bewildering array, as the focus in biology turned towards genes and other molecules. The functionality, nevertheless, lies at the level of the channels themselves and their interactions in the cell as a whole. This realisation played a large role in my development towards systems biological interpretations of the problems we were working on. While I was happy to admire the detailed molecular structures of the channel proteins as they became revealed, I also knew that all the main cardiac *functions* of these channels were revealed in the work described in this chapter and in Chapters 1 and 5.

References

Beeler, G. W. and Reuter, H. (1977) 'Reconstruction of the action potential of ventricular myocardial fibres', *Journal of Physiology*, 268, pp. 177–210.

Burdon-Sanderson, J. and Page, F. J. M. (1883) 'On the electrical phenomena of the excitatory process in the heart of the frog and of the tortoise, as investigated photographically', *Journal of Physiology*, 4, pp. 327–338.

Deck, K. A. and Trautwein, W. (1964) 'Ionic currents in cardiac excitation', *Pflügers Archiv, European Journal of Physiology*, 280, pp. 65–80.

Dudel, J., Peper, K., Rudel, R. and Trautwein, W. (1967) 'The potassium component of membrane current in Purkinje fibres', *Pflügers Archiv, European Journal of Physiology*, 296, pp. 308–327.

Hecht, H. H., Hutter, O. F. and Lywood, D. W. (1964) 'Voltage current relations of short Purkinje fibres in sodium-deficient solution', *Journal of Physiology*, 138, pp. 5–7.

Luo, C.-H. and Rudy, Y. (1991) 'A model of the ventricular cardiac action potential – depolarization, repolarisation and their interaction', *Circulation Research*, 68, pp. 1501–1526.

Luo, C. and Rudy, Y. (1994a) 'A dynamic model of the cardiac ventricular action potential: I. simulations of ionic currents and concentration changes', *Circulation Research*, 74, pp. 1071–1097.

Luo, C. and Rudy, Y. (1994b) 'A dynamic model of the cardiac ventricular action potential: II. Afterdepolarizations, triggered activity and potentiation', *Circulation Research*, 74, pp. 1097–1113.

McAllister, R. E. and Noble, D. (1966) 'The time and voltage dependence of the slow outward current in cardiac Purkinje fibres', *Journal of Physiology*, 186, pp. 632–662.

McAllister, R. E. and Noble, D. (1967) 'The effect of subthreshold potentials on the membrane current on cardiac Purkinje fibres', *Journal of Physiology*, 190, pp. 381–387.

McAllister, R. E., Noble, D. and Tsien, R. W. (1975) 'Reconstruction of the electrical activity of cardiac Purkinje fibres', *Journal of Physiology*, 251, pp. 1–59.

Noble, D. (1966) 'The Biological origins of self', *Common Factor*, 4, pp. 24–31.

Noble, D. (1967a) 'Charles Taylor on teleological explanation', *Analysis*, 27, pp. 96–103.

Noble, D. (1967b) 'The conceptualist view of teleology', *Analysis*, 28, pp. 62–63.

Noble, D. (2008) 'The Birth of The House of God', In *Return to The House of God: Medical Resident Education, 1978–2008*, (ed. M. Kohn and C. Donley), pp. 1–8.

Noble, D. and Stein, R. B. (1966) 'The threshold conditions for initiation of action potentials by excitable cells', *Journal of Physiology*, 187, pp. 129–162.

Noble, D. and Tsien, R. W. (1968) 'The kinetics and rectifier properties of the slow potassium current in cardiac Purkinje fibres', *Journal of Physiology*, 195, pp. 185–214.

Noble, D. and Tsien, R. W. (1969a) 'Outward membrane currents activated in the plateau range of potentials in cardiac Purkinje fibres', *Journal of Physiology*, 200, pp. 205–231.

Noble, D. and Tsien, R. W. (1969b) 'Reconstruction of the repolarization process in cardiac Purkinje fibres based on voltage clamp measurements of the membrane current', *Journal of Physiology*, 200, pp. 233–254.

Sanguinetti, M. C. and Jurkiewicz, N. K. (1990) 'Two components of cardiac delayed rectifier K+ current. Differential sensitivity to block by class III antiarrhythmic agents', *Journal of General Physiology*, 96, pp. 195–215.

Tsien, R. W., Giles, W. R. and Greengard, P. (1972) 'Cyclic AMP mediates the action of adrenaline on the action potential plateau of cardiac purkinje fibres', *Nature, New Biology*, 240, pp. 181–183.

Tsien, R. W. and Noble, D. (1969) 'A transition state theory approach to the kinetics of conductance changes in excitable membranes', *Journal of Membrane Biology*, 1, pp. 248–273.

Weidmann, S. (1951) 'Effect of current flow on the membrane potential of cardiac muscle', *Journal of Physiology*, 115, pp. 227–236.

Chapter 3

Analytical Mathematics
of Excitable Cells*

3.1. Introduction

The mathematical lectures of Dr Few at UCL (Chapter 1) must have left a lasting impression on me. My first reason for attending them was the entirely practical necessity of convincing people to let me have time on the Mercury computer. That strategy was successful. But his lectures also rekindled a deep respect for mathematics itself. There is something immensely satisfactory in finding an analytical solution to a problem. Computational models are necessary, of course, but from a mathematical viewpoint, they can be seen as just high-level number crunching. They can create excitement, certainly, for example, when they are used to clarify a counterintuitive result, as in the apparent permeability changes in Chapter 1 (Fig. 1.9) and the pacemaker pulse comparison in Chapter 2 (Fig. 2.5). But the satisfaction that comes from deriving closed-form analytical solutions is much deeper for a very simple reason: *such solutions are general*. The story of the Bessel function solutions that solved the challenge to the existence of thresholds for repolarisation in Chapter 1 illustrates this point. It was not just a particular computation that resolved the issue. That might have been seen as a special case. Rather, it was the general conclusion resulting from the maths itself. Any point excitation in a two- or three-dimensional network will decay so rapidly that even gross non-linearities in membrane properties get hidden.

---

*Readers without mathematical skills might choose to skip this chapter.

That is the power of an analytical solution to a mathematical problem. It was decisive. Point polarisation of a tissue was never used again in current or voltage control of the heart. An elegant method had been destroyed by a single ugly mathematical fact. Experimentalists sometimes forget that it can be that way round too.

This kind of generality is also what is required for the future of the systems approach to biology. I will return to that question at the end of this chapter and again in Chapter 9.

3.2. Current Flow in Excitable Cells

One of the reasons why most biological models are based on computation using differential equations is that this approach can deal with almost any degree of non-linearity in the system being studied. By contrast, analytical solutions are easiest to obtain in linear systems. A good example of this distinction in excitable cell theory is that of cable theory, which is required to study the conduction of the electrical impulse in the heart and other muscles, as well as in nerves. For voltage ranges in which the system is reasonably linear, which is usually the sub-threshold region, analytical solutions in the form of error functions (Hodgkin and Rushton, 1946) can be obtained for the spread of electrical current. By contrast, once the excitation threshold is reached, the linear approximations are no longer valid. The conduction of the impulse requires non-linear differential equations to be solved. But must this always be by numerical analysis? Or are there ways in which closed-form solutions can be obtained?

Frustrated by the limits of the relatively slow computers of those days, this was the question that I was tackling with Peter McNaughton in the early 1970s. Peter is now professor and head of the Department of Pharmacology in Cambridge. The background to this particular project was a collaboration with Julian Jack and Dick Tsien. Since the arrival of Julian Jack in the department in 1964, he and I had organised a Saturday morning seminar on excitation theory. It was restricted to Saturday mornings since most faculty members were not at all sure that this degree of mathematics was required in a department of physiology, though there were exceptions, notably Brian Lloyd, a respiratory physiologist.[1] We

---

[1]Brian Lloyd, with Dan Cunninghasm, was the father of mathematical respiratory physiology. His equations for control of respiration by O_2 and CO_2 are still used today.

started the course with work done with Dick Tsien on a book on the mathematics of excitation theory (Jack *et al.*, 1975). That book was planned in the late 1960s, but it was taking us far longer to write than we expected. My work on the project had reached Chapter 12, the last chapter apart from a mathematical appendix, and was entitled 'Non-Linear Cable Theory: Analytical Approaches Using Polynomial Models'.

As this title says, the idea was to represent the non-linearity of the ion channel current by polynomial functions. Any continuous function can be fitted to varying degrees of accuracy with polynomials, the accuracy being determined by the order of the polynomial. For functions of interest in electrophysiology, these start with third-order polynomials, which can generate the kind of current–voltage relations often seen in excitable cells, with three crossing points (see the continuous curve in Fig. 3.1).

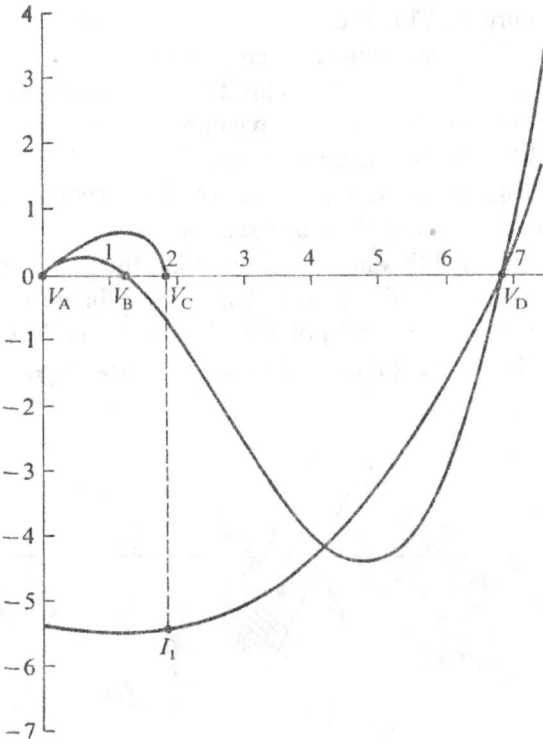

Figure 3.1. Third-order polynomial representation of the idealised current–voltage relation of an excitable cell. See text for description.

Source: Adapted from Fig. 12.1 and equation (12.12) of Jack *et al.*, 1975.

The lower (V_A) represents the resting state. The upper one (V_D) represents the voltage towards which excitation carries the potential. The middle one (V_B) is a threshold, beyond which excitation produced by a uniform current applied to the membrane is all or nothing.

Examples for fitting of such polynomials to experimental data are to be found in Fig. 5 of Jack *et al.*, 1975. A cubic polynomial can be a very close fit to data from squid nerve, investigated in Hodgkin & Huxley's work. A quintic (fifth order) works well for a cardiac Purkinje fibre.

My reaction to the use of polynomial functions in this context was to realise that we had a tool for the analytical investigation of thresholds in excitable cells. The derivation of the threshold V_C in Fig. 3.1 is an example of this approach. The curve leading up to it is the current that would have to be applied in the point excitation of a uniform cable. This requires a larger voltage change since a longer section of the cell needs to be generating inward current. This is called the liminal length (see Noble (1972) for an analysis of this parameter). There is a simple relationship between the threshold V_C and the form of the current–voltage relation, which is that the integrals of the positive and negative current areas should be equal, as shown in Fig. 3.2, which is from a review of the Hodgkin–Huxley equations that I wrote some years earlier (Noble, 1966) (reproduced as Fig. 9.1 in *Electric Current Flow in Excitable Cells*).

I immediately used this approach to revisit the repolarisation threshold problem discussed in Chapter 1. The results for uniform excitation, point excitation of a cable and point excitation of a sheet are shown in Fig. 3.3 (from Fig. 12.2 of *Electric Current Flow in Excitable Cells*).

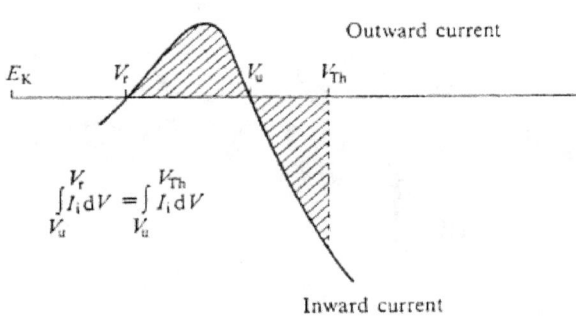

Figure 3.2. Relation of the voltage threshold for cable excitation (V_{Th}) to the membrane current–voltage relation. V_r is the resting potential and V_u is the voltage threshold for excitation by uniform polarization. V_{Th} is given by the point at which the two integrals are equal.

Source: Adapted from Noble, 1966.

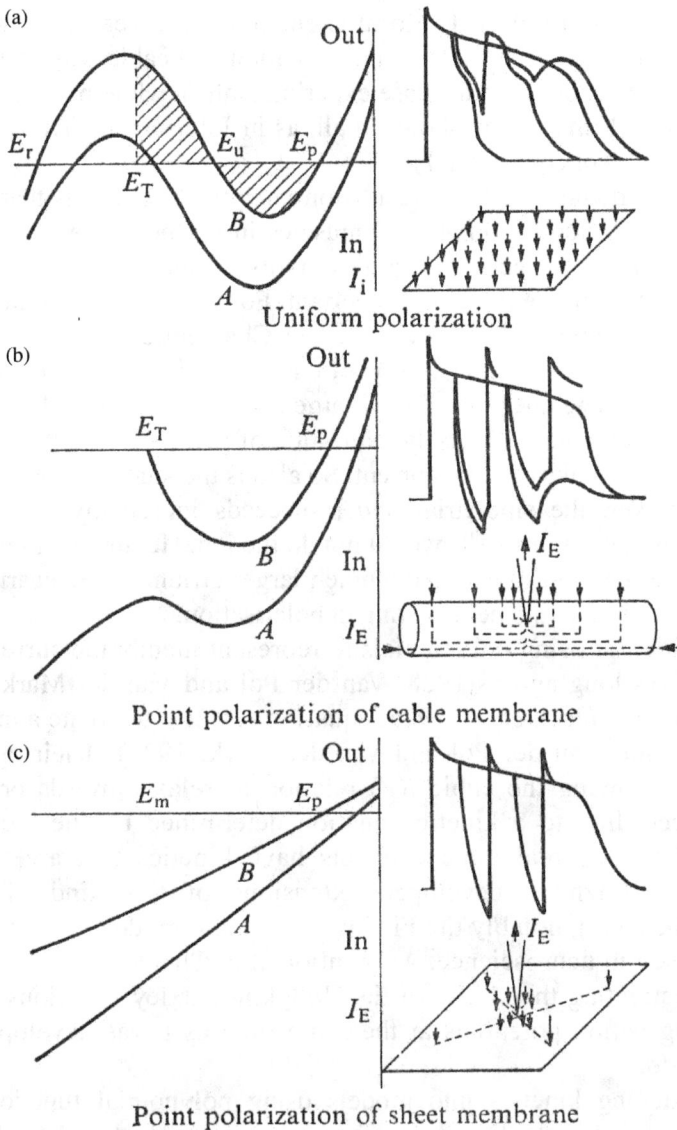

Figure 3.3. Effect of tissue and stimulus geometry on the repolarisation threshold in cardiac muscle. In each case, the early and late $i(V)$ relations are shown on the left and the expected voltage changes with time on the right. (a) Uniform polarization: The thresholds always lie between the resting and plateau levels and approach the plateau as repolarisation occurs. (b) Point polarization of a cable: The early threshold then disappears. The late threshold is shifted strongly in a negative direction. (c) Point polarization of a two-dimensional sheet: The current–voltage relations become linearized and thresholds disappear.

Source: Adapted from Jack *et al.*, 1975.

The predictions are clear. Uniform excitation can reveal the true membrane ionic current thresholds, point excitation of a cable will lead to very negative thresholds (as is the case experimentally), while point excitation of a sheet will find no thresholds at all, as in Johnson and Tille's cardiac experimental work (Chapter 1).

The significance of these results on analytical models is that natural forms of excitation of nerves and muscles in the body are not uniform. The geometry of the cells and tissues involved, and of the origins of the excitatory stimuli, are therefore important. For example, one of the mechanisms of arrhythmia to be discussed in Chapters 5 and 7 reflects the existence of a focus where a region of tissue is ischemic or shows a similar pathology. The question of how large such a region must be to excite the heart as a whole is precisely a question of geometry and the strengths of sources and sinks of ionic current. So also is the study of how the natural pacemaker, the sino-atrial node, succeeds in reliably exciting the whole heart. Just a few cells would not do the job. Hundreds of thousands of cells are required to excite the much larger atrium of the heart, which has a powerful i_{K1} channel resisting depolarisation.

The idea of using polynomials to represent membrane currents was not new. As long ago as 1928, Van der Pol and Van der Mark used a cubic equation (Jack *et al.*, 1975, equation (11.21)) to create a model of heart rhythm (Van der Pol and Van der Mark, 1928). Their approach involved allowing the cubic $i(V)$ relation to relax upwards or downwards according to a kinetic equation determined by the membrane potential just as real ionic channels have kinetics that are voltage-dependent. Fitzhugh developed extensions of this kind of model (Fitzhugh, 1961), notably the Fitzhugh–Nagumo model which has been widely used in neuroscience. As mentioned in Chapter 1, Fitzhugh was also investigating the ability of the Hodgkin–Huxley equations to generate long action potentials at the same time as I was developing the 1962 model.

Introducing kinetics into models using polynomial functions was therefore already established. Our interest was in whether these kinetics could be made to model more closely the actual Hodgkin–Huxley kinetics, including delayed activation.

Peter McNaughton and I were poring over some pages of mathematical scribble when Peter Hunter joined us to look at what we were doing. He saw one of the solutions immediately and that was what led us to invite him to join us on the paper. I did not realise then that this would be the first of many interactions with Peter Hunter, leading eventually to the Physiome Project.

As will be clear from the articles referenced in this book, I have interacted with many mathematicians, and mathematically inclined engineers and physicists. I find that they belong to (at least) two classes: those who have to work hard through the developments and proofs (a class to which I certainly belong) and those to whom maths is a natural language of thought, who see solutions before they even begin to prove them. The same is true of the language of music, where also I belong to the world of the slow plodders. But belonging even to the slow stream enables one to appreciate the work of those who swim naturally in the medium. I have had the pleasure of knowing and working with some brilliant mathematicians and musicians. The significance of the interaction with musicians will become clear in Chapter 9. Let's now return to the mathematicians.

3.3.   Equations for the Speed of Conduction

The problem Peter McNaughton, Peter Hunter and I were staring at on those sheets of paper comes from the fact that voltage-gated ion channels do not switch on or off immediately when the voltage is changed. They take time to do so. The simplest kinetics leads to this occurring in a first-order process, with an exponential time course, like the potassium channels described in Chapter 2. So, we have to deal not only with the non-linearity but also with the gating reaction itself. The exponential case could be treated analytically, which in itself was a major advance (an insight also seen at the same time by Alan Hodgkin). But, as I said, this is just the simplest case. The sodium channels in nerves and muscles usually display an initial delay in activation; the time course is sigmoid rather than exponential. In the Hodgkin–Huxley equations, this is represented by supposing that more than one gate is involved and that all have to be opened for the conduction of ions to occur.

The question therefore was whether analytical solutions could be obtained even for these, more usual, cases. The answer was *yes*, and that is one of the major contributions of the paper by Hunter *et al.* (1975). It was only improved on many years later in the work of Rob Hinch in my lab (Hinch, 2002).

The result for the case where three gates are involved (the standard Hodgkin–Huxley sodium channel activation process) allows the following equation to be derived for the conduction velocity of the impulse:

$$\theta = k_1(a/2R)^{1/2} \cdot (k_2g)^{1/8} \cdot C^{(-5/8)}$$

where θ is the velocity, a is the fibre radius, R is the longitudinal resistance inside the fibre, g is the maximal channel conductance, C is the membrane capacitance, and the k's are constants.

Some parts of this result had been known for a long time. That the speed of conduction increases as the square root of fibre size and inversely with internal resistance (first term) was well known. So also was the dependence on membrane capacitance (third term). The novelty lies in the extraordinarily small dependence on the ion channel conductance, g. For delayed activation involving three gates (m particles in the Hodgkin–Huxley equations), the velocity increases only as the eighth root of the conductance.

These findings were fully incorporated into *Electric Current Flow in Excitable Cells*. In fact, so far as non-linear cable theory is concerned, this was one of the major contributions of that book, which grew out of an initial desire simply to teach the mathematics of nerve and muscle excitation to students, but then developed into a set of major original contributions to the field. It remains the standard text and is now freely available as a complete PDF file on the website of *The Music of Life:* https://www.denisnoble.com/wp-content/uploads/2021/01/JNT-2.pdf.

The equation for the conduction velocity has many consequences for the biology of the transmission of impulses. Consider, first, the need for an organism to transmit information rapidly. The faster it knows about a challenge in the environment, and the faster it can react to that in its own motion, the more likely it is to survive. One might think that the best solution to this problem is to tell the genome to make more sodium channels and pack them into the cell membrane as much as possible. This would increase g in the equation. But, after an initial gain in speed as channels are added, the gain becomes very small (Fig. 3.4). To double the speed would require 256 (2^8) times as many channels, an increase of more than two orders of magnitude! It would be much more effective to simply increase the fibre size, where a doubling of speed can be achieved with just a fourfold increase in radius. This is precisely what has happened in evolution. The most spectacular example in invertebrates is the squid

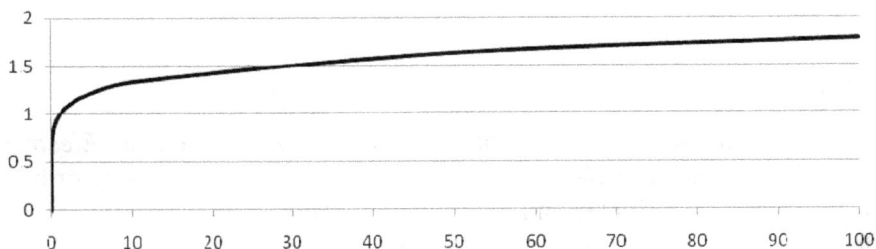

Figure 3.4. Graph showing how an eighth root increases with its argument. Most of the increase occurs at very small values. Between 10 and 100, in the increase of an order of magnitude, the eighth root increases by only 33%. This gain in speed could be achieved by just an 80% increase in fibre size.

giant axon responsible for the rapid escape that the squid achieves when it activates its huge mantle muscle to jet-propel itself away from a predator. It also ejects a cloud of ink to confuse the predator. This was the nerve fibre discovered in the squid by Young in 1937, which formed the basis of the Nobel-Prize-winning work of Hodgkin and Huxley in 1952. I have often reflected that the prize might reasonably have been shared with Young whose discovery and understanding of the function of large nerve fibres in invertebrates was so important.

The solution for high conduction velocity found in vertebrates is to enclose the majority of the length of the nerve in an insulating myelin sheath with only occasional nodes acting as transmission stations with dense sodium channels. Some of the mathematics of conduction in myelinated nerves is dealt with in Chapter 10 of *Electric Current Flow in Excitable Cells*.

Not only would packing more sodium channels along the entire fibre length produce only a very modest gain in speed, but eventually it actually *reduces* the speed. The reason is that the gating process in channels involves transient movement of charge, which adds to the capacitance of the membrane. As the equation shows, adding capacitance would slow the speed much faster than the channel conductance increases it (compare the 5/8th power for C with the 1/8th power for g). Pages 128–129 of the paper by Hunter *et al.* (1975) deal with equations for gating capacitance. As Hodgkin also realised, this means that there is an optimal channel density. In order for cells to achieve this, some form of electro-transcription coupling must occur.

3.4. Conclusions

3.4.1. *Pluses*

I believe that the work with Julian Jack and Dick Tsien on *Electric Current Flow in Excitable Cells*, together with the developments described in the paper with Hunter and McNaughton (Hunter *et al.*, 1975) reproduced for this chapter, form a high-water mark in mathematical physiology. Given the difficulty and density of the mathematics, it was reasonable to expect only very modest sales. It surprised me therefore that the book was sufficiently appreciated to come out as a second edition in paperback form in 1983. A fact of which the authors are immensely proud is that the corrections amounted to just one bracket in one equation, a tribute to the careful proofreading and the help of some of our students and colleagues. The book remains the standard text half a century later and I am delighted that OUP gave permission for the PDF to be freely available at https://www.denisnoble.com/wp-content/uploads/2021/01/JNT-2.pdf.

3.4.2. *Minuses*

Like a beached whale, the book and the articles appeared during the period when the tsunami of molecular biology was sweeping away almost everything in its path. Classical physiology took a back seat during those years. Moreover, despite the paperback publication, the number of physiologists who could understand the book and the papers on which it was based was small. The contribution of mathematics to physiology is still a vexed question. I suspect that the innumerable copies of *Electric Current Flow in Excitable Cells* to be found on physiologists' bookshelves was, at least in part, a showing-off for visitors! How many really understood it is a different question.

3.4.3. *Contributions to systems biology*

It may be, however, that the recent growth of systems biology will help change that situation. Systems biology itself could benefit from the lessons of our work on the mathematics of excitation. Not everything of value in the quantitative analysis of biological function has to be cranked out as differential equations solved numerically on fast computers. On the contrary, and as the *general* nature of the solutions to non-linear problems in excitation theory shows, greater understanding can be achieved when closed-form analytical results are obtained. That is possible sometimes even in highly non-linear systems.

The insight into the existence of an optimal density of ion channels is important. The electro-transcription coupling that must occur to ensure this is also a form of 'downward causation', to which I will return in Chapter 9.

References

Fitzhugh, R. (1961) 'Impulses and physiological states in theoretical models of nerve membrane', *Biophysical Journal*, 1, pp. 445–466.

Hinch, R. (2002) 'An analytical study of the physiology and pathology of the propagation of cardiac action potentials', *Progress in Biophysics and Molecular Biology*, 78, pp. 45–81.

Hodgkin, A. L. and Rushton, W. A. H. (1946) 'The electrical constants of a crustacean nerve fibre', *Proceedings of the Royal Society B*, 133, pp. 444–479.

Hunter, P. J., McNaughton, P. A. and Noble, D. (1975) 'Analytical models of propagation in excitable cells', *Progress in Biophysics and Molecular Biology*, 30, pp. 99–144.

Jack, J. B., Noble, D. and Tsien, R. W. (1975) *Electric Current in Excitable Cells*. Oxford: Oxford University Press.

Noble, D. (1966) 'Applications of the Hodgkin-Huxley equations to excitable tissues', *Physiological Reviews*, 46, pp. 1–50.

Noble, D. (1972) 'The relation of Rushton's "liminal length" for excitation to the resting and active conductances of excitable cells', *Journal of Physiology*, 226, pp. 573–591.

Van der Pol, B. and Van der Mark, J. (1928) 'The heartbeat considered as a relaxation oscillation and an electrical model of the heart', *Philosophical Magazine Supplement*, 6, pp. 763–775.

Chapter 4

Insight into the T Wave of the Electrocardiogram

4.1. Historical Background

Electrophysiology may be the area of physiology that most students avoid, because of its mathematical difficulty, but it is also one of the most useful in clinical medicine. The reason is that it is an area where practising doctors need to be familiar with the standard way of monitoring the function of the heart, which is to detect arrhythmias, periods when the heartbeat ceases to be regular. Diagnosing when this may have serious consequences is important. That was also clear to physiologists in the 19th century when understanding the significance of electrical recordings of the heart as detected by surface body recordings began to be understood. Almost everyone nowadays is familiar with electrocardiograms (ECG or EKG) as a tool of diagnosis. That story goes back to a predecessor as a professor of physiology at Oxford, John Burdon Sanderson.

In fact, John Burdon Sanderson was the *first* professor of physiology at Oxford University. In moving from UCL to Oxford, I was following in his footsteps since he was also at UCL before Oxford. There seems to be a tradition of UCL handing heart physiologists over to Oxford. But the similarities do not stop there. Sanderson (I will come to his double surname later) was the great-grandfather of the cardiac action potential. He was the first to show its very long duration.

To assess the significance and technical achievement of this work, we need some imagination. In the 1880s, there were no computers, of course, other than Charles Babbage's pioneering work on his 'difference engine' – but that never worked in his lifetime. Moreover, there were no electronic recording systems of any kind. The only instrument available was the capillary electrometer, which uses a mercury column that can move because of changes in surface tension when an electrical signal is applied to it. Sanderson showed great ingenuity. He shone a light beam at the column, and the resulting shadow of the column was projected onto a photographic apparatus. There were no microelectrodes that could be employed to access the interior of a cell. The connection to the intracellular potential of the heart muscle was achieved by placing one of the electrodes on a damaged region of the heart.

The result was a remarkably accurate recording of the shape and duration of the cardiac action potential. Moreover, he demonstrated quite clearly that one of the waves, called the T wave, of the electrocardiogram corresponds to the repolarisation phase of the action potential. The classical paper in which all this is described is that of Burdon-Sanderson and Page (1883). Notice the double-barrelled surname.

I naturally wondered what caused him to change. Could the change have been influenced by the alphabetical order of names insisted on in those days by the *Journal of Physiology*? I checked the early minute books of the Physiological Society. All the early references to him, including his own signature, are to John Sanderson or to John B Sanderson. I suspect therefore that he was also faced with the problem of the alphabetical order. Dick Tsien (see Chapter 2) should have been called Dick Chien, one of the alternative transliterations of his family name, used by another Californian scientist with the same Chinese surname, Shu Chien! He would then, rightfully, have been the first author on the many papers we published together.

4.2. Repeat of the Classic Experiment

Burdon-Sanderson did this work on the tortoise and the frog. In both cases, the relationship between the action potential and the surface electrocardiogram was very simple, as shown in Fig. 4.1, which was a repeat of his experiment that was done in my laboratory, also using injury to gain access to the intracellular potentials, and which was used to illustrate my book *The Initiation of the Heartbeat* (Noble, 1975, Fig. 1.5).

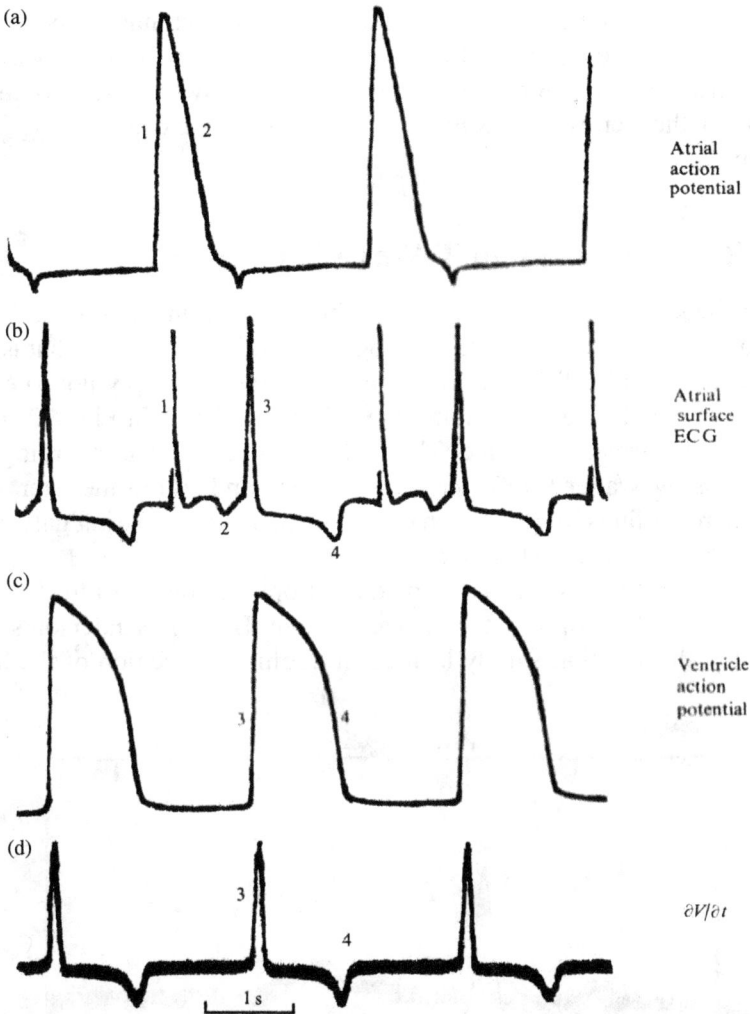

Figure 4.1. Simultaneous recordings of atrial (a) and ventricular (c) action potentials, the electrocardiogram (b) and the first derivative of the ventricular action potential (d) made in a tortoise with the injury method originally used by Burdon-Sanderson and Page (1883) but using a modern pen recorder instead of a capillary electrometer.

Source: Adapted from Noble, 1975, Fig. 1.5.

Notice that each deflection in the electrocardiogram (labelled 1, 2, 3 and 4) is related to a large change in the corresponding action potential. Each action potential generates a rapid positive wave at its beginning and a slower negative wave at its end. This is what we would expect since the current that flows externally should, to a first approximation, be proportional to the first derivative of the action potential.[1] The first derivative of the ventricular action potential is shown in the lowest trace; its resemblance to the ventricular waves (3 and 4) in the electrocardiogram is obvious.

4.3. The Mammalian T Wave is Positive

So far, the story seems simple. One might even imagine that what doctors can do when they see an electrocardiogram is to infer directly what is happening in the cells of the heart. In mammals, that is simply not true. The T wave of the electrocardiogram (wave 4 in Fig. 4.1, T in Fig. 4.2) is not negative. It is usually positive! The first recordings of electrocardiograms in humans by Waller (1887) showed this fact, and so did the later, more accurate recordings by Einthoven (1913) using the faster string galvanometer that he introduced (Fig. 4.2).

Why that is the case has occupied electrophysiologists of the heart for many years. An important clue came from Burdon-Sanderson's own work. He showed that simply heating or cooling one region of the heart,

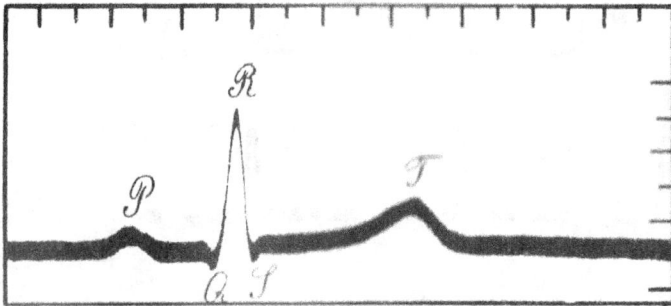

Figure 4.2. Electrocardiogram recorded between the surfaces of the two arms in a man using a string galvanometer. Time in 50 ms intervals, potential in 0.5 mV intervals. The labelling of the waves, P, Q, R, S and T is in Einthoven's own hand.

Source: Reproduced from Einthoven, 1913.

---

[1] Strictly speaking, this should be the first derivative of the *spatial* variation. In a uniformly propagating impulse, the time derivative is proportional to the spatial derivative.

which has the effect of shortening or lengthening the action potentials in that region, can invert the T wave or change its shape. It will now be obvious why, in 1984, when the British Heart Foundation offered me one of its prestigious chairs, I chose for it to be named after Burdon-Sanderson as the most natural tribute to a great predecessor.

What the heating experiments showed is that differences in action potential duration, however they might be caused, could have profound effects on the T wave. This is why the T wave is perturbed during cardiac ischemia (reduced blood flow). As I showed in the work reviewed in Chapter 1, hyperkalaemia (excess extracellular potassium – one of the effects of ischemia) alters the action potential duration, and these changes can explain the T wave effects during ischemic heart attacks.

But why is the normal T wave positive in mammals? The most natural explanation would be that, even in the normal heart, there are natural variations in action potential duration between different regions. This is the case. There are variations across the ventricular wall and between the base and apex of the ventricle that follow a rule: the regions of the heart tissue that are excited first repolarise last. This is illustrated diagrammatically in Fig. 4.3.

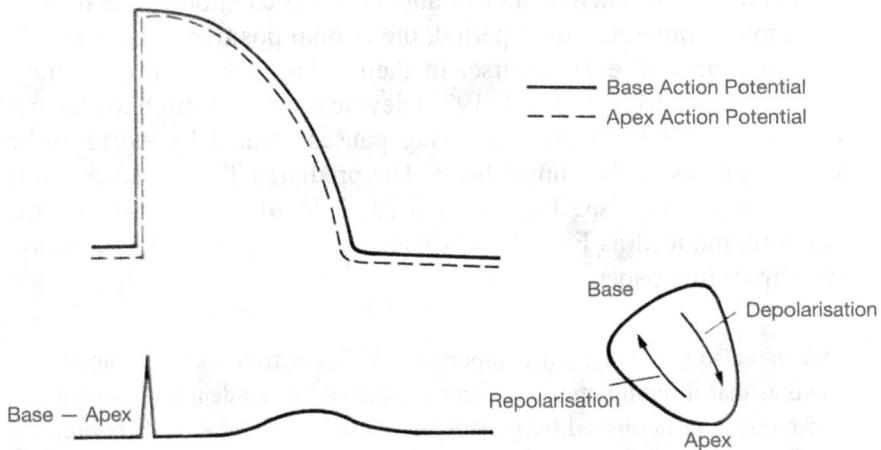

Figure 4.3. Mechanism by which action potential duration differences can give rise to a positive T wave. This diagram shows the difference between the base and apex. The electrocardiogram can then be represented, to a first approximation, as the difference between these, which is always positive.

Source: Adapted from Noble, 1975.

4.4. The 1976 *Nature* Paper

The question that remains to be answered is what generates the action potential differences that are responsible for the direction and shape of the T wave. The paper reprinted for this chapter reveals that, surprisingly, these differences are not present in quiescent tissue. They develop rapidly during repetitive activity. This was shown by cutting isolated tissue from different regions of the ventricle, allowing them to rest for a period of time (30–60 secs) and then stimulating them repetitively. The first action potentials in the train are not significantly different. The second and sub-sequent action potentials all show the difference (see Cohen *et al.*, 1976).

To unravel what is happening, we used two interventions, both expected to reduce the activity of the sodium–potassium exchange pump: cooling and application of the drug ouabain, which specifically inhibits the sodium–potassium exchanger. In both cases, the duration differences were reduced or abolished.

What might cause these effects? In the article, we speculate that activity-dependent changes in ion concentrations may be involved, but we were not able to test this idea experimentally.

How do these findings correlate with what is known to happen to the whole heart in situ? There is a clear and simple prediction. If the heart is allowed to rest quiescent for a period, the normal positive T wave should change in form and even reverse. In their exhaustive book on clinical observations, Scherf and Schott (1973) devote a whole section to changes in the shape of the T wave, following pauses created by extrasystolic (additional) beats of the human heart. The predicted T wave inversion is seen in many cases (see Figs 3.22, 3.23, 3.24 of their book, and the example reproduced as Fig. 19 in Noble and Cohen (1978)). It is worth quoting from that paper:

> We can also make one other important prediction from our hypothesis. This is that if the duration gradients are activity dependent they should be reduced or abolished by periods of quiescence. We have, therefore, looked carefully at the electrocardiographic literature for examples of this. One of the most important papers on this subject was published by Scherf (1944). He investigated the electrocardiogram during beats fol-lowing the pause that usually occurs following a premature extrasystole. He found that in some, though not in all, cases the T wave is greatly changed in the post-extrasystolic beat. Scherf correctly noted that this

phenomenon, which is seen in about 30% of cases, is directly related to the long diastole. Ashman *et al.* (1945) also reached this conclusion. It does not depend on the abnormality of the previous ventricular extrasystole. This is shown by the fact that it also occurs following atrial extrasystoles and following long diastolic intervals in patients with atrial fibrillation and A-V block.

4.5. Subsequent Developments

This venture into the interpretation of the T wave of the electrocardiogram has had far-reaching practical consequences. Failures in the process of repolarisation and changes in the T wave are critically important in the pharmaceutical industry. Failure of repolarisation is one of the causes of cardiac arrhythmias, sometimes even fatal. The T wave is one of the easiest measurements that can be made non-invasively to monitor the repolarisation process. Many failures of drugs through side effects on the heart have shown T wave changes as the first warning sign. It is therefore used as a biomarker for potential adverse effects on the heart. The work of my team, and that of many others around the world, has now used mathematical modelling to help the industry detect the possibility of these problems at early stages in the drug development process.

These were the reasons why I found myself lodged in 1997 in the Watergate Hotel in Washington, waiting to give evidence at a hearing of the Food and Drug Administration (FDA). My host was the pharmaceutical giant, Hoffmann-La Roche.

A few months earlier, I had been contacted by Jean-Paul Clozel, who was leading the preclinical assessment of a new Roche compound, mibefradil (posicor). To understand the significance of this compound, I need to explain that just as potassium channels have been found to come in various types, so also have calcium channels. In the case of the heart, there are two main calcium channels. The L-type channels (L for long-lasting since they switch off slowly) form the largest group and they are responsible for triggering contraction. The T-type channels (T for transient) are activated at lower voltages and so play a role in pacemaker activity (Lei *et al.*, 1998). In the ventricle though, they play a very minor role. Blocking these channels does little either to the action potential or to contraction. By contrast, they play a major role in arterial smooth muscle, so blocking them can relax these muscles and lower blood pressure.

A specific T-type calcium channel blocker could therefore be a useful drug, helping to lower blood pressure without adversely affecting the mechanical power of the heart. Mibefradil had the distinction of being the first compound to have such an effect. Naturally, there was considerable excitement about this development and whole symposia were devoted to T-type channels and the actions of mibefradil (Tsien *et al.*, 1998).

Jean-Paul's problem was that mibefradil had been found to modify the shape and duration of the T wave (Fig. 4.4). For the FDA, this was a bad sign. The T wave was spread out over a longer period of time and even broken up into several waves. Yet, the action potential was not prolonged by the usual criteria. How could this be? The situation was confusing because the usual criteria for measuring the timing of repolarisation from the T wave could not be applied. How could one measure the peak time of the T wave if it sometimes had more than one peak? There was no way in which any experiment could be conceived at that time that could even begin to answer the question. Could modelling help? That was the question Jean-Paul put to me. With Rai Winslow and Peter Hunter (see Chapter 7), we were already incorporating cellular models of the heart into tissue and

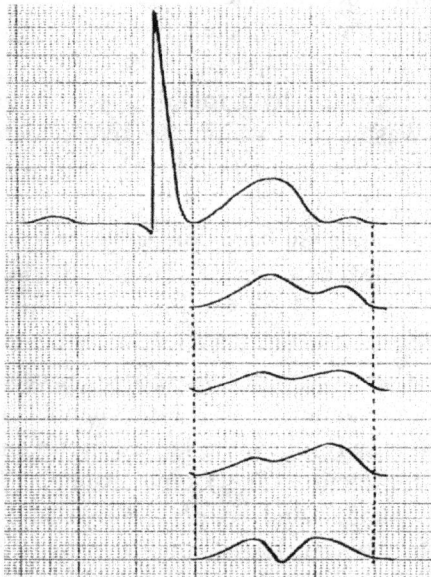

Figure 4.4.   Examples of some of the strange T wave patterns produced by the T-channel blocker, mibefradil.

organ models using large parallel computers. We could even reproduce a version of the electrocardiogram. Could we therefore demonstrate that, at the least, the action of mibefradil on the T wave of the electrocardiogram could be understood and that it was not necessarily a bad sign? Rai and I set to work. In fact, as I remember it, Rai set aside around 6 weeks of time and work on the Johns Hopkins parallel supercomputer to work flat out over the Christmas and New Year period to do the calculations. They worked. It was possible to show that, given the effects that mibefradil was known to have at the molecular and cellular level, the T wave changes were, in principle, understandable. In particular, a broadening of the T wave could be consistent with a shortening of the action potential.

I need to be very clear here about what the modelling was showing. Understanding was the key, not prediction. And that understanding meant that the T wave changes did not necessarily mean that a hidden electrophysiological effect was lurking somewhere that could itself be arrhythmogenic. It could not be interpreted to mean that mibefradil was entirely safe since we were modelling only the heart, and only one aspect of heart activity. As it turned out, mibefradil did subsequently have to be withdrawn because of metabolic interactions in the liver. One would need to have good models of that organ also to deal with those problems computationally. I will return to this issue in Chapter 7.

Sometime before the FDA hearing, Roche organised a meeting in New York to assess the evidence. I was asked to attend at very short notice, but had to decline because of a prior commitment that meant I could not travel to the US in time. That problem was solved very quickly. Roche agreed to fly me out via Concorde, the only time I have had the opportunity to fly in that remarkable aircraft.[2] I left after breakfast in London to arrive *before* breakfast in New York. I still have a Concorde cut glass paperweight on my desk as a reminder of that extraordinary experience. People said that you could see the curvature of the earth because Concorde flew at such a high altitude. Certainly, I saw curvature of the horizon, but I am not sure that it can be interpreted as sufficient evidence that the earth is a sphere. Even a flat earth would have a horizon and, from any given vantage point, it would seem to meet the sky at the same distance from the observer in all directions, so creating a curved horizon. I think one has to be much higher up to see that the earth really is a sphere.

---

[2] Remarkable for speed and cruising height, of course. The fuel consumption was unacceptably high.

The International Space Station has now solved that problem. From that height above the earth, it is clearly a rotating sphere.

The FDA hearing was also an extraordinary experience. Unlike Europe, these are held in full public view. There is almost an element of theatre in the proceedings. Big issues are at stake. The investment in a single drug can represent more than $1 billion. That experience has been put to good use in my subsequent interactions with the pharmaceutical industry to try to find a solution to the T wave problem. The reason is that I was deeply impressed by one of the FDA officials, Ray Lipicky. He was the official who directly interrogated me before the Committee was given the chance (Noble, 1997). But, more relevantly, he had spent enormous amounts of time checking the electrocardiograms himself. I think he came to the conclusion that, from those alone, you could conclude very little. I agree. The T wave is a seriously flawed biomarker.

The reason goes deeper than the T wave itself. As originally shown by Burdon-Sanderson, the T wave reflects the repolarisation phase of the cardiac action potential. That is also an unreliable biomarker on its own. Certainly, many drugs that delay repolarisation cause arrhythmia, but some do not. It all depends on what other actions they have. These complexities need to be understood, not ignored.

Many of my subsequent publications have been concerned with this question (Noble, 1984; Noble and Cohen, 1978; Taggart *et al.*, 1979) and the related one of developing and assessing drugs that may avoid the problem (Fink *et al.*, 2009; Fink and Noble, 2010; Noble, 2008; Noble and Noble, 2006; Noble and Varghese, 1998; Noble and Noble, 2000; Rodriguez *et al.*, 2006). This project grew into a European-funded consortium called PreDiCT (http://www.vph-predict.eu/) in collaboration with many pharmaceutical companies to help in developing better biomarkers. Ray Lipicky was one of its Scientific Advisors, while Roche (the developer of the first T-channel blocker – the focus of that 1997 hearing at the FDA) was one of the members of the consortium.

4.6. Conclusions

4.6.1. *Pluses*

So, far as I am aware, the main conclusions of the *Nature* paper on the T wave and the subsequent 1978 article in *Cardiovascular Research* with Ira Cohen are both correct and have not been improved on.

4.6.2. *Minuses*

Yet, no one seems to have followed up on the lead showing the activity dependence of the T wave.

4.6.3. *Contributions to systems biology*

The subsequent developments leading to the PreDiCT project clearly form an example of the application of systems biology to clinical problems and in the pharmaceutical industry. In fact, the systems approach has attracted considerable interest from clinical and pharmaceutical research, as shown by two articles we published in a special issue of the Nature group journal *Clinical Pharmacology and Therapeutics* (Kohl *et al.*, 2010; Rodriguez *et al.*, 2010).

References

Ashman, R., Ferguson, F. P., Gremillion, A. L. and Byer, E. (1945) 'The effect of cycle length changes upon the form and amplitude of the T deflection of the electrocardiogram', *American Journal of Physiology*, 143, p. 453.

Burdon-Sanderson, J. and Page, F. J. M. (1883) 'On the electrical phenomena of the excitatory process in the heart of the frog and of the tortoise, as investigated photographically', *Journal of Physiology*, 4, pp. 327–338.

Cohen, I ., Giles, W. R. and Noble, D. (1976) 'A cellular basis for the T wave of the electrocardiogram', *Nature*, 262, pp. 657–661.

Einthoven, W. (1913) 'Uber die Deutung des Elektrokardiograms', *Pflügers Archiv, European Journal of Physiology*, 149, pp. 65–86.

Fink, M. and Noble, D. (2010) 'Pharmacodynamic effects in the cardiovascular system: The modeller's view', *Basic & Clinical Pharmacology & Toxicology*, 106, pp. 243–249.

Kohl, P., Crampin, E., Quinn, T. A. and Noble, D. (2010) 'Systems biology: An approach', *Clinical Pharmacology and Therapeutics*, 88, pp. 25–33.

Lei, M., Brown, H. F. and Noble, D. (1998) 'What role do T-type calcium channels play in cardiac pacemaker activity?' In *Low-Voltage-Activated T-type Calcium Channels* (ed. R. W. Tsien, J.-P. Clozel and J. Nargeot), pp. 103–109. Chester: Adis International.

Noble, D. (1975) *The Initiation of the Heartbeat*. Oxford: Oxford University Press.

Noble, D. (1984) 'The T wave of the electrocardiogram in relation to intracellular action potentials', In *Proceedings of 3rd Einthoven Symposium on Past and Present Cardiology* (ed. A. C. Arntzenius, A. J. Dunning and H. A. Snellen), pp. 34–42. Leiden: Spruyt.

Noble, D. (1997) 'Transcript NDA 20-689', In *80th Meeting of Cardiovascular and Renal Drugs Advisory Committee.* Washington: FDA.

Noble, D. (2008) 'Computational models of the heart and their use in assessing the actions of drugs', *Journal of Pharmacological Sciences*, 107, pp. 107–117.

Noble, D. and Cohen, I. (1978) 'The interpretation of the T wave of the electrocardiogram', *Cardiovascular Research*, 12, pp. 13–27.

Noble, D. and Noble, P. J. (2006) 'Late sodium current in the pathophysiology of cardiovascular disease: consequences of sodium-calcium overload', *Heart*, 92, pp. iv1–iv5.

Noble, D. and Varghese, A. (1998) 'Modeling of sodium-calcium overload arrhythmias and their suppression', *Canadian Journal of Cardiology*, 14, pp. 97–100.

Noble, P. J. and Noble, D. (2000) 'Reconstruction of the cellular mechanisms of cardiac arrhythmias triggered by early after-depolarizations', *Japanese Journal of Electrocardiology*, 20 (Suppl 3), pp. 15–19.

Rodriguez, B., Burrage, K., Gavaghan, D., Grau, V., Kohl, P. and Noble, D. (2010) 'Cardiac applications of the systems biology approach to drug development', *Clinical Pharmacology and Therapeutics*, 88, pp. 130–134.

Rodriguez, B., Trayanova, N. and Noble, D. (2006) 'Modeling cardiac ischemia', *Annals of the New York Academy of Sciences*, 1080, pp. 395–414.

Scherf, D. (1944) 'Alterations in the form of the T waves with changes in the heart rate', *American Heart Journal*, 28, p. 332.

Scherf, D. and Schott, A. (1973) *Extrasystoles and Allied Arrhythmias.* London: Heinemann.

Taggart, P., Carruthers, M., Joseph, S., Kelly, H. B., Marcomichelakis, J., Noble, D., O'Neill, G. and Somerville, W. (1979) 'Electrocardiographic changes resembling myocardial ischaemia in asymptomatic men with normal coronary arteriograms', *British Heart Journal*, 41, pp. 214–225.

Tsien, R. W., Clozel, J.-P. and Nargeot, J. (ed.) (1998) *Low-Voltage-Activated T-type Calcium Channels.* Chester: Adis International.

Waller, A. D. (1887) 'A demonstration on man of electromotive changes accompanying the heart's beat', *Journal of Physiology*, 8, pp. 229–234.

Chapter 5

The Surprising Heart

5.1. A Telephone Call from Milan

> He telephoned me from Milano in January 1980 to tell me this result
> and the same night I was able to use a computer program he and I had
> developed together to show that his new interpretation of i_{K2} as a non-
> specific inward current i_f could give a full and accurate theoretical
> account of the i_{K2} results. (Noble, 1984, p. 10)

'He' was Dario DiFrancesco; 'this result' was to cause turmoil for
several years. And it was itself the outcome of turmoil, including that in
my own laboratory.

5.2. Ion Concentration Changes

The turmoil began with the discovery that some components of electrical
current change recorded from multicellular heart tissue arise from changes
in the concentrations of sodium, potassium and calcium ions rather than
just from the opening and closing of protein channels, in the presence of
fixed concentrations inside and outside the cells.

In Chapter 2, I referred to the fact that the analysis of the slow potassium ion channels that we now call i_{Kr} and i_{Ks} had also revealed some
'minor exceptions' in the ionic current traces. These exceptions were even
slower components that Dick Tsien and I were reluctant to attribute
directly to ion channel activity. They were so slow that they could not have
influenced the heart excitations much from beat to beat, so we preferred

the idea that they may have resulted from slow changes in potassium ion concentrations, perhaps in the spaces between the heart cells. My wife, Susan Noble, worked on these changes in frog atrial muscle, described first in her 1972 doctoral thesis and in later papers (Brown *et al.*, 1976a, 1976b; Noble, 1976).

These developments revealed a difficult mathematical dilemma. The analysis of the kinetics and voltage dependence of ionic channels depends on knowing where the electrical and chemical energy gradients balance each other at what is called the reversal potential. If the relevant ion concentrations are changing, then so also is this potential. How could one possibly separate changes attributable to that process from the real kinetics of the channels? Some critics of our work said it was impossible and that the results obtained from voltage clamp work on multicellular tissue of the heart were, quite simply, an inextricable mess. In fact, Johnson and Lieberman (1971)[1] wrote a long review saying precisely that, even describing the frog atrial muscle preparation voltage clamped by the double sucrose gap method as an 'insanitary preparation'. The cardiac electrophysiology sessions at the 1977 World Congress of the International Union of Physiological Sciences, in Paris, and a satellite meeting in Poitiers were dominated by this controversy. It didn't help that I gave my plenary lecture at that Congress in French (this was the last IUPS Congress to be multilingual).[2]

Later, we worked on the equations for this kind of problem. Using the maths of perturbation theory, DiFrancesco and I were able to show that it is possible to dissect out the gating kinetics from other components (DiFrancesco and Noble, 1980b), while Susan Noble and Wayne Giles used Provencher's (1976) DISCRETE program to confirm the accuracy of her hand analysis of the multi-exponential changes seen experimentally. These two mathematical analyses formed the background of the work that ensured that I had a computer program ready for that fateful telephone call.

---

[1] Yes, this is the same Johnson as in the Johnson & Title article described in Chapter 1.

[2] An American wag said I must have done that to confuse my critics, who didn't think to get earphones to listen to the simultaneous translation. It was, of course, done, as I nearly always do in Francophone countries, as a courtesy. Nor did it help that the slide projector light bulb blew in the middle of the lecture, so that I had to continue for about 15 minutes without notes or slides. It was the first occasion on which I had lectured in French in free conversational form. A Russian lecturer following me said (in English) that he could now believe that pigs could fly!

5.3. Discovery of the 'Funny' Current (Figure 5.1)

But before we come to what that call revealed, there is another important discovery to note. In the 1970s, we developed a method to study induced pacemaker rhythm in strips of frog atrial muscle (Brown *et al.,* 1976a, 1976b). When Wayne Giles came to Oxford for two years, we extended this work to spontaneously beating strips of frog sinus venosus (the natural pacemaker of the heart). When the electrical potential was made very negative, we recorded a slowly developing inward current (also observed by Seyama (1976) in rabbits at about the same time). We called it the 'additional current', without at that time realising its importance.

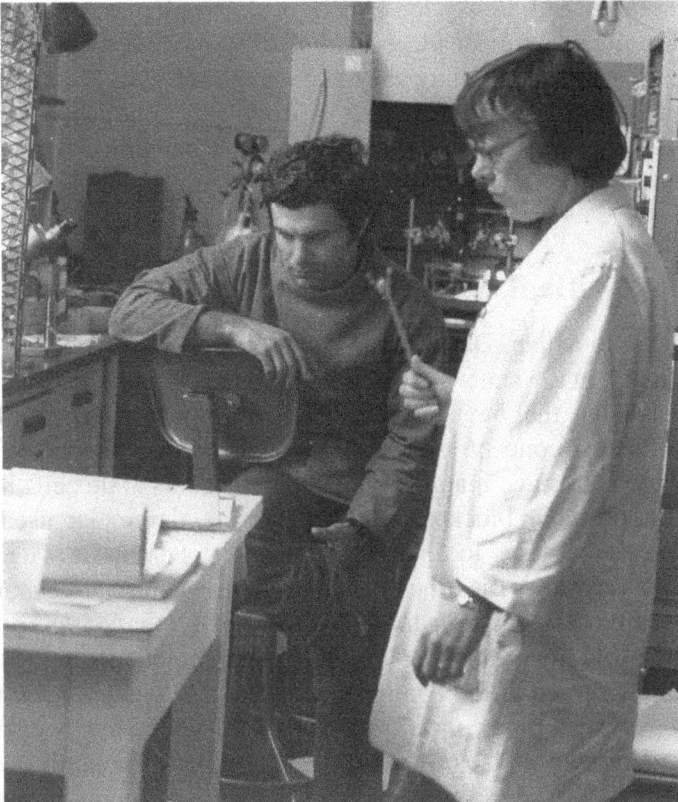

Figure 5.1. 'What a funny current!' Dario DiFrancesco and Hilary Brown contemplate an experimental problem in the lab. It might have been the first recordings of the 'funny' current.

When Dario DiFrancesco (Figure 5.1) came from Cambridge to join our group in 1977, he was keen to investigate pacemaking in mammalian tissue and to use the so-called 'small preparation' of rabbit sinoatrial node tissue recently pioneered by Aki Noma and Hiroshi Irisawa (Noma and Irisawa, 1976). It was technically challenging to use this preparation so that the tiny 200-micron-diameter ball of tissue continued its spontaneous beating and it was extremely difficult to impale it with two microelectrodes and obtain a uniform control of voltage. Dario was the lead experimenter in this work and his persistence and skill were rewarded by recordings of a remarkable (as it seemed to us at the time) inward current which appeared in the potential range of the pacemaker depolarisation, precisely the same range as i_{K2} in the Purkinje fibres that I had worked on with Dick Tsien. But, unlike i_{k2}, *it did not show reversal at the potassium equilibrium potential.* Instead, it continued to increase even beyond that potential. 'There's that funny current' we would say. So, i_f it became (Brown and DiFrancesco, 1980). A paper they published in *Nature* showed that it was reversibly increased by adrenaline (Brown *et al.,* 1979), contributing to the acceleration of the heart rhythm by adrenaline.

5.4. The Critical Experiment

On his return to Milan, Dario carried out a critical experiment. Reluctant to accept that i_f in the sinus node and i_{K2} in Purkinje tissue really were two different channels that happened to share a lot of characteristics, he wondered whether ion concentration changes could account for the difference. The Purkinje cells have many i_{K1} channels; the sinus node cells have few or none. Suppose one blocked i_{K1} in the Purkinje tissue? He used barium ions, which were known to do this. The result was dramatic. The reversal potential that had identified the channel as a pure potassium channel simply disappeared! *The Purkinje tissue then resembled that of the sinus node.*

After he told me of this result by telephone (no emails in those days), I immediately turned to the computer program that we had been using to analyse the effects of ion concentration changes. With a few tweaks, it was ready to address an audacious question. The reversal potential results in Purkinje fibres looked clean, and the dependence on potassium ion concentration followed the Nernst equation faithfully. Similar results had also been obtained by Shrier and Clay (1982) using embryonic chick hearts. In fact, their reversal potential recordings were even cleaner than ours.

5.5. Mapping the Two Theories

The audacious question was how could this possibly arise if the channels were really not pure potassium ion channels. Could nature have set a cruel trap for electrophysiologists? This was an even greater challenge for computational modelling than anything I had tackled this far. The question to be settled was this: could the current variations attributable to ion concentration changes have kinetics so similar to the channel kinetics that they could cancel each other out cleanly? And not only do that, they also had to do so in a way that created an accurate illusion of a Nernstian reversal. 20 years on from 1960, the computers were much faster. I didn't need weeks of calculations. By the next morning, I was able to tell Dario that it had all worked out like a dream. If one replaced i_{K2} in the McAllister–Noble–Tsien (1975) model with a mixed (sodium and potassium) current, i_f, like that in the sinus node, and included the accumulation and depletion of potassium ions in the spaces between the cells, the resulting mixture behaved just like i_{K2} (see Fig. 5 of 'The Surprising Heart' (Noble, 1984)). Not only did this explain the Nernstian behaviour of the reversal potential, which is now recognised to be false, but it also explained why the current disappeared when one removed sodium ions (Chapter 2), and why the apparent reversal potential was always a few mV negative to the expected reversal. We set to work to analyse this initially very strange result mathematically and published it the same year (DiFrancesco and Noble, 1980a). The full details were published two years later (DiFrancesco and Noble, 1982).

If you are not an ion channel electrophysiologist, it is hard to appreciate the full nature of the shock this result produced. The Nernst equation was, after all, the gold standard for identifying the ionic composition of a channel current. Yet, we had shown that it could 'lie'. So unbelievable was the result that I had many rounds of correspondence with those who had also identified the i_{K2} mechanism in other species and tissues of the heart. It took some time for the significance to sink in. Was it just a coincidence? If so, why should it occur so widely? In fact, the mathematics we did showed that it was far from a coincidence, and it only required very moderate (10%) changes in ion concentration to produce the effects. Yet again in the work of my laboratory, not only was mathematical modelling necessary to reveal the relevant insights but mathematical *analysis* was also required; the apparently (and initially) obscure work on perturbation theory had borne fruit, just as the use of Bessel functions had done (Chapter 1). Once again, analytical mathematics had complemented computation to produce

Figure 5.2. With Dario DiFrancesco (centre) after he won the Grand Prix Lefoulon-Delalande in Paris in 2008. To the left is Professor Alain Carpentier, President of the Committee that chooses the laureate. The prize was awarded for his discovery of the i_f channel. I declare a conflict of interest here: I am a member of the Committee charged with making this choice. Of course, for that reason, I did not participate in that award decision.

results of complete generality, which could provide powerful explanations of counterintuitive experimental results.

When Dario (Figure 5.2) and I wrote the 1982 article, we looked for an appropriate piece of literature that could reflect the painful yet joyful nature of this journey of discovery, and which would also reflect his and my interests. I had already been exploring the medieval Troubadour poets, and had found that Dante Alighieri had praised one of the Troubadour poets, Arnaut Daniel (circa 1180), in the *Purgatorio* of *La Divina Commedia* (circa 1308–1321). Not only did he praise Arnaut as the best craftsman (il miglior fabbro) of poetry in the language of the people, in his case Occitan, he also wrote these verses of his great work in Occitan rather than Italian as his tribute. The verse fully expresses the pain of discovery, yet how easy it is for others to follow where the discoverer has led:

Ara vos prec, per aquella valor
que vos guida al som de l'escalina,
sovenha vos a temps de ma dolor
(Purg., XXVI, 140–147)

It can be translated thus:

Therefore do I implore you, by that power
Which guides you to the summit of the stairs,
Remember my suffering, in the right time

The stairs could, of course, be the stairs of the mountain to Paradise in the *Paradiso*. I rather like to think of them as the difficulty the researcher experiences in blazing a path up Mount Discovery. Others coming later can climb readily what he found difficult. 30 years later, no one today finds the i_f story difficult at all. But its transformation from the apparently secure i_{K2} story was far from easy.

5.6. The DiFrancesco–Noble (1985) Model

The obvious next step was to develop the McAllister–Noble–Tsien 1975 model to replace i_{K2} by i_f. But that was much easier said than done. It took a full 5 years of development. The reason was that it was not just a matter of replacing one ionic channel mechanism with another. It also involved modelling global ion concentration changes for the first time[3] in an electrophysiological model of the heart, including intracellular calcium signalling. Dario and I did that because it was necessary to explore fully what we had discovered. We did not know then that we would be creating the seminal model from which virtually all subsequent cardiac cell models would be developed. There are now over 100 such models for various

---

[3]Beeler and Reuter (1977) had incorporated changes in calcium concentration in a sub-space, but they did not incorporate the ion exchangers and pumps that were required in our work.

parts of the heart and many different species to be found (downloadable) on the CellML website (www.cellml.org)

Extending biological models is often like tumbling a row of dominoes. Once one has fallen, many others do too. The reason is that all models are necessarily partial representations of reality. The influences of the parts that are not modelled must either be assumed to be negligible or to be represented, invisibly as it were, in the assumed boundary conditions and other fixed parameters of the model. Once one of those boundaries is removed (and fixed concentration was one of those boundaries in previous models), by extending out to a different boundary, other boundaries become deformed too. In this case, modelling external potassium changes required modelling of the influence of those changes not only on the ion channels already in the model but also on exchange mechanisms, like Na-K-ATPase (sodium pump) and the Na-Ca exchanger. That in turn required the model to extend to modelling internal sodium concentration changes, in turn requiring modelling of intracellular calcium changes, which then required modelling of the sarcoplasmic reticulum uptake and release mechanisms. For a year or two, it was hard to know where to stop, where to stake out the new boundaries.

It was this process that uncovered a major insight. To simplify the story, I will focus on just one of the new mechanisms: the sodium–calcium exchanger. This had been discovered in the heart by Harald Reuter (Reuter and Seitz, 1969), who also discovered cardiac calcium currents (Reuter, 1967) (see Chapter 2). Harald and I complemented each other experimentally, with his focus being on calcium ion transport, while mine had been on potassium ion transport. We first met at a course in Homburg, Germany, in 1966, where I was invited to lecture on electrophysiological theory (the basis of what became *Electric Current Flow in Excitable Cells* (Chapter 3)). Harald was a student of the course. As such, he certainly benefited from it much more than I did. To go on to discover both the calcium current and the sodium–calcium exchanger was phenomenal. I shall never understand why this was not recognised with a major international award.

Until the work with Dario DiFrancesco, Harald and I had kept off each other's territory. But now Dario and I were forced to enter his area. But in doing so, we were also forced to change it. Harald's work strongly suggested that the sodium–calcium exchanger was electrically neutral. To extrude one calcium ion carrying two positive charges, the exchanger would transport two sodium ions into the cell. It was using the sodium gradient to maintain the calcium gradient and doing so in an assumed 2:1 ratio. Sodium ions carry one charge, so the charge balance would be neutral.

We put this ratio into the model. *It didn't work.* Instead of driving the intracellular calcium down to around 100 nM, a level at which the mechanics of the cell would be quiescent, it barely achieved 10 times that level, i.e. around 1 μM, at which level the cell would be in a permanent state of contraction! Clearly, during each heartbeat, the levels of intracellular calcium must oscillate between these extremes, but not be permanently at the high end of the range. This discovery forced us to abandon the assumption of neutrality. But that was to fly in the face of the best experimental evidence at that time, i.e. Harald Reuter's pioneering work.

Finding some equations to do so was not difficult. Lorin Mullins (1981) had already explored this possibility mathematically. So, we adopted his equations and tried out a 4:1 ratio (his favoured one). That also did not work. It drove the calcium levels far too low. Obviously, 3:1 was the right choice. That is what we went for, and subsequent experimental work has fully confirmed that choice.

It is hard to tell a complex story like this one without initially telling a lie to simplify it. I will now correct the lie. It is not true to say that we had *no* experimental leads to favour an electrogenic activity of the exchanger. We had the experimental evidence in our own laboratory in the form of electrical currents recorded by a Japanese student, Junko Kimura, working with Hilary Brown, Susan Noble and Anne Taupignon. They were finding an extra component of inward current in the sinus node of the heart (Brown *et al.*, 1983). Per Arlock in Sweden had found a similar component (Arlock and Noble, 1985). The problem was that these results were controversial and regarded with suspicion. Perhaps they were yet another example of Johnson and Lieberman's (1971) strictures against our work: were they just a set of artefacts? If these results were correct, then not all the slow components of inward current in the heart were attributable to calcium channels carrying calcium into the cells to activate contraction; some were also attributable to sodium–calcium exchange driving calcium out of the heart as part of the process of relaxation.

I think there is a general rule for revolutionaries in experimental physiology. You are allowed one revolution at a time. If the substitution of i_f for i_{K2} had been the only major change in the DiFrancesco–Noble model, then it might have passed the critics. The evidence for it rapidly became uncontroversial and strong. Since we were forced by the logic of

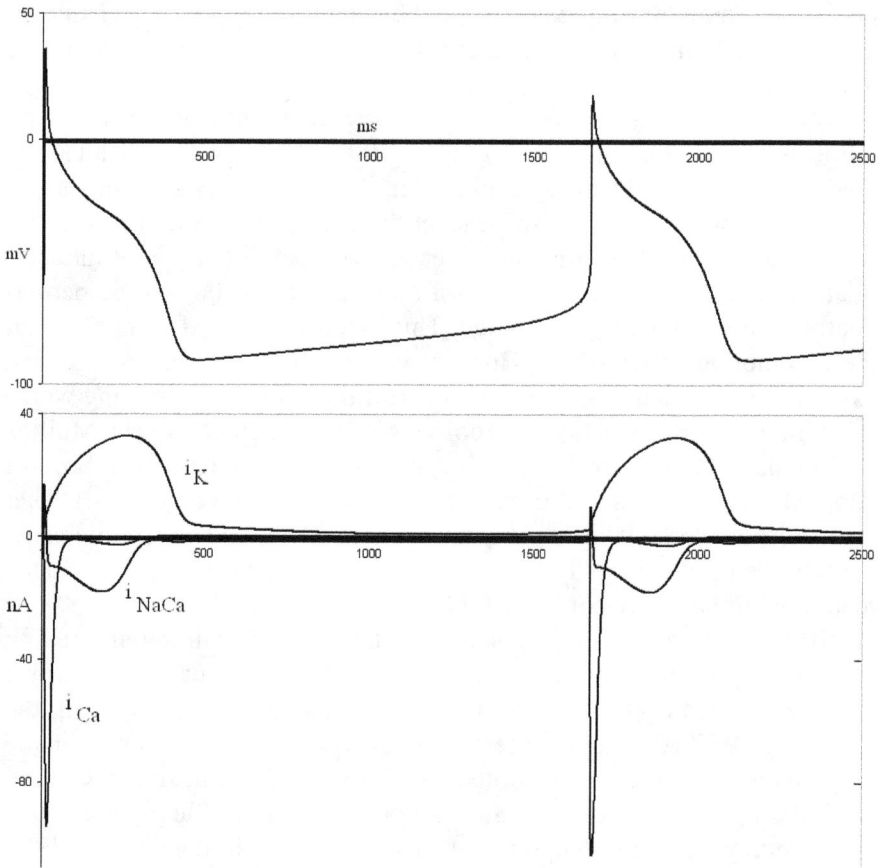

Figure 5.3.   The DiFrancesco–Noble 1985 Model. This figure was made using COR and the cellML coding of the model. It shows the temporal relationship between activation of the L-type calcium current and the almost immediately following activation of the sodium–calcium exchange current.

the new modelling to ask people to accept a second revolution at the same time, for which the evidence was much weaker, the critics thought we were going too far out on a limb with our speculations. A set of papers, including the DiFrancesco–Noble model, but also including all the experimental work of Hilary Brown, Junko Kimura, Susan Noble and Anne Taupignon (Brown *et al.*, 1984a, 1984b, 1984c; Noble and Noble, 1984), was sent off to the *Journal of Physiology*. They were rejected as a set, with the modelling paper singled out as unsuitable. This is the reason why that set of papers, and the DiFrancesco–Noble model itself (Figure 5.3), was published in journals of the Royal Society instead. We were very reluctant to split up the experimental and theoretical work.

There was, however, a curious twist to the story. Coincidentally, I had been asked to give The Physiological Society's Annual Review Lecture that year. This is the lecture that became 'The Surprising Heart' (Noble, 1984), which incidentally not only contains all the new heresies, but fully explains the reasons for them. So, the major results were published in the *Journal of Physiology* after all, albeit in review form.

Heresies? 25 years later, it is very hard to see what all the fuss was about. Almost everything in the DiFrancesco–Noble model is now part of the standard story, and it forms the canonical model on which all subsequent ones in many laboratories around the world have been based.

In fact, the current generated by sodium–calcium exchange during and following the action potential has been used as an indicator of intracellular calcium changes both in our laboratory, for example, in work done with Trevor Powell, Terry Egan, Jean-Yves LeGuennec, David Fedida, Yakhin Shimoni and many others (Egan *et al.,* 1989; Fedida *et al.,* 1987; Noble *et al.,* 1991; Noble and Varghese, 1998), and in other laboratories around the world. Here in the UK, David Eisner designed an accurate method for estimating the calcium content of the sarcoplasmic reticulum, from which he has developed an elegantly complete account of calcium cycling in cardiac cells (Eisner and Sipido, 2004; Eisner *et al.,* 2000; Venetucci *et al.,* 2007). This method is now widely used.

5.7. Sodium–Calcium Exchange

As for the second heresy, that of attributing components of inward current to the sodium–calcium exchanger, that rapidly became confirmed in a spectacular way by experiments performed by Junko Kimura after she returned to Japan.

She returned to work at the National Institute of Physiological Sciences in Okazaki. I think she must have reported on the work she did with us in Oxford, to be met by a considerable degree of scepticism from her mentor, Akinori Noma. But they did exactly the right thing, which was to put the whole story to a rigorous experimental test. They eliminated virtually all currents other than the exchanger and measured its voltage and ion dependence (Kimura *et al.*, 1986, 1987). The test was simple. If the exchanger was electrogenic, there had to be a detectable current. I could hardly believe the stroke of luck we had. Not only did they record the expected current, the comparison between the theoretical and experimental results was remarkable and speaks for itself. Rarely does theoretical speculation benefit from this degree of good fortune in subsequent experimental work (Fig. 5.4).

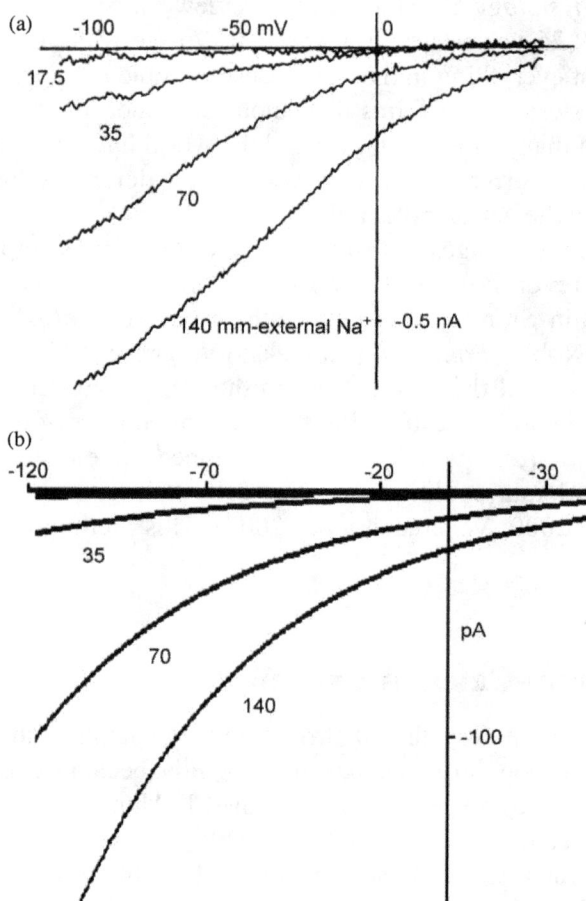

Figure 5.4.   Comparison between the experimental results (a) on the sodium–calcium exchange current (Kimura *et al.*, 1987) and those given by the equations (b) used in the DiFrancesco–Noble 1985 model. The curves were obtained at different levels of external sodium ions between 17 and 140 mM.

The main difference between the experiment and theory is that the increase in negative current at very negative potentials is not as great in the experiments. Something clearly limits the speed of the exchanger at these extremes. This difference hardly affects the results obtained in the normal range of electrical activity of the heart, which lies between -90 and $+30$ mV.

5.8. Sodium–Calcium Exchange in Sinus Node Rhythm

Rhythm in the natural pacemaker of the heart, the sinus node, differs considerably from that in Purkinje tissue. The main inward current involved in the upstroke of the action potential is not the sodium current. It is the calcium current. Hilary Brown and Susan Noble, in the Oxford laboratory, were exploring this mechanism and the role of the i_f channel, together with their collaborators, Dario DiFrancesco, Junko Kimura and Anne Taupignon. Some of the results found their way, discreetly, into the bottom drawer, thought initially to be unpublishable, even perhaps artefactual as I noted earlier in this chapter. These were the strange, slow, transient components of inward current that followed the activation of the calcium channels.

The results eventually came out of the bottom drawer to see the light of day in published papers because they became the focus of another fruitful interaction between the experimental and computational work. The theoretical work predicted not only that a slow transient inward current *should* follow the more rapid calcium current, as in Fig. 5.3, but also that, under the right conditions, it could even *precede* it. This was precisely what those bottom-drawer experimental results showed!

The critical experiment was quite clever. Allow the pacemaker depolarisation in the nodal cells to approach, but not go beyond, the threshold for the L-type calcium current. Then, hold the voltage at that level. In many experiments in which this protocol was used during the last third of the pacemaker depolarisation, a slow transient inward current could be recorded that looked just like the one that usually follows activation of the calcium channel (Fig. 5.4).

By that time, Susan Noble and I had modified the DiFrancesco–Noble Purkinje model to create the first model of a sinus node cell (Noble and Noble, 1984). So, we applied the same protocol to the derived model. The results showed the same behaviour as in the experiments. This is shown

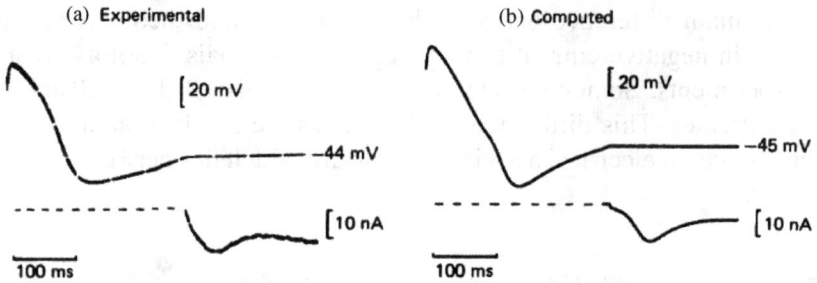

Figure 5.5. (a) Experimental result. The membrane potential in the sinus node is allowed to follow the natural time course of the pacemaker depolarisation until −44 mV is reached. The voltage is then clamped at this point. A very slow inward current develops. The activation and turn-off take almost equal periods of time, which is characteristic of the slow inward currents attributable to sodium–calcium exchange. (b) Reconstruction of this type of current record using the sinus node version (Noble and Noble, 1984) of the DiFrancesco–Noble (1984) equations.

Source: Adapted from Brown *et al.*, 1983, 1984a.

in Fig. 11 of 'The Surprising Heart', which was taken from the Noble–Noble 1984 paper (see Fig. 5.5). What is happening here? The slow transient current is attributable to calcium release from the sarcoplasmic reticulum. The calcium signal then activates current flow through the sodium–calcium exchanger.

This insight raises the possibility that there are at least two oscillator mechanisms in the natural pacemaker of the heart, one driven primarily by membrane ion channels, the other by an internal calcium oscillator. This is the idea that was later followed up in great detail by Ed Lakatta's group in the US. An issue of *Circulation Research* focuses on this question and on how the two oscillators entrain each other (Christoffels *et al.*, 2010; DiFrancesco, 2010; Efimov *et al.*, 2010; Lakatta *et al.*, 2010; Noble *et al.*, 2010; O'Rourke, 2010). As we will see in the next chapter (Chapter 6), the sinus node pacemaker is robustly backed up by several mechanisms that can substitute for each other.

5.9. Sodium–Calcium Exchange in Cardiac Arrhythmias

After leaving my laboratory, Dick Tsien moved to Yale. One of his first experiments in his new laboratory followed up on some uninterpreted squiggles of ionic current that he had found in the Oxford results. With

Jon Lederer (Lederer and Tsien, 1976), he demonstrated that these transients could be regularly evoked by treating cardiac cells with ouabain, digitalis and similar steroids that are known to block the sodium pump. It didn't take long for us to realise that the transient inward currents we were observing in 1984 were of the same kind. In fact, Lederer and Tsien had already proposed sodium–calcium exchange as one of the possible mechanisms. The idea was simply waiting for proof that the exchanger could be electrogenic.

This work has formed the basis of unravelling a major cause of heart arrhythmias. What a block of the sodium pump achieves is a rise in intracellular sodium as less sodium is pumped out of the cells. Such a rise also occurs in many pathological conditions, including ischemia. The mechanism is now well established. As intracellular sodium increases, the sodium gradient used by the sodium–calcium exchanger becomes weaker. Less calcium is pumped out of the cell, leading to a rise in intracellular calcium. If this rise is sufficient to trigger release of further calcium from the sarcoplasmic reticulum, then the calcium oscillator can begin cycling either transiently or indefinitely. Each calcium release generates an inward current through activation of sodium–calcium exchange, which can then trigger full excitation. This is one mechanism of what are called ectopic beats in the heart (see Chapter 7).

Some of the models developed from the DiFrancesco–Noble model reproduce this mechanism. I will return to this question after relating the next story, which was also started by a telephone call.

5.10. A Telephone Call from Los Angeles

This call was from Don Hilgemann (Fig. 5.6) working in Los Angeles. It was early in 1985.

Before I explain what Don was excited about and how I reacted, I need to describe the political situation in the UK at that time. We were in the middle of a ferociously strong and determined Thatcher government. With virtually no opposition (she had completely crushed the opposition in the 1983 general election), her government was pursuing a severe financial policy that, amongst many other things, seriously cut the budgets for research. I was a member of a Research Council Grants Committee at that time and experienced directly the chaos this provoked. Senior scientists in the UK were driven to despair by the situation in which we

Figure 5.6.   With Junko Kimura (left) and Don Hilgemann (right) at the international meeting on sodium–calcium exchange in Baltimore in April 1991.

found ourselves. We sometimes did not know how pitifully small the allocation would be until the grants committees met. By then, it was too late to adjust our marking, and think of a situation in which we could probably save only 1 out of maybe 20 grant applications. To say we were shocked is an understatement. The heart was being eaten out of the glory that was British science in its 20th century heyday. The period of international recognition in the form of Nobel and similar prizes was threatened. And, indeed, the frequency of those honours did fall quite dramatically during the next two decades.[4]

---

[4]The average number of UK Nobel Prizes in science per decade over the period of the 50s, 60s, and 70s was 11. The average over the 80s and 90s dropped to 5.5. During the first decade of the 21st century, it is back up at 11. It is hard to interpret the figures since there must be delays between funding and results, and many other factors are involved. Over the hundred years of the Nobel Prizes, the UK has performed very well when the number is expressed per capita of population.

These were the circumstances in which the council of my university had put forward a proposal, normally uncontroversial, to award the Prime Minister an honorary degree. Whether wisely or with foolhardiness, I and many others said 'no'. In Oxford, such a proposal requires ratification by a full meeting of academics known as Congregation. I made one of the major speeches opposing the degree, which was rejected by an overwhelming majority, perhaps the only setback that the Prime Minister experienced during that period.

There was joy throughout the academic community in the UK. But whether it was a wise decision was another matter. There was also, and naturally, fury and anger from her supporters and from the government. By the time the DiFrancesco–Noble 1985 paper was published in the *Philosophical Transactions of The Royal Society*, I was a national television personality, appearing regularly on behalf of the science and academic communities. This initiative eventually became the pressure group, Save British Science, in January 1986 (now a highly effective organisation known as the Campaign for Science and Engineering). It was launched by the majority of Britain's Nobel laureates, over 100 Fellows of the Royal Society and over 1500 fellow scientists from all over the UK.

This is not the place to go into the pros and cons of that instinctive reaction to Oxford University Council's proposal. One of my former colleagues from UCL simply commented, 'What you did was just; whether it was wise is another question'. I agreed. That was the choice, between justice and wisdom, difficult though it was.

The telephone call from Don Hilgemann came in the middle of the hectic round of media appearances. I simply told him I was too busy to talk. His call must have been squeezed in somewhere between the BBC and *The Times*.

Fortunately, he persisted. He telephoned back later and said very firmly, 'You have got to listen to me'. He was greatly excited by the DiFrancesco–Noble paper since he had experimental results that it could explain. He had been measuring the changes in extracellular calcium in the heart during each beat (Hilgemann, 1986a, 1986b). His results showed that the calcium level between the cells does indeed fall quickly (though not by much – these were difficult experiments) as calcium enters the cells through the calcium channels. But within only 20 milliseconds, it was being pumped out again! The thinking up to that time was that restoration of the ionic gradients may occur over a much longer period, perhaps after relaxation of the heart. Don recently wrote to me to give his version of the thinking:

> It seemed abundantly clear when I finished my PhD in 1981 that calcium was leaving the myocyte during the contraction cycle, not between beats. The question was how and with what electrical implications. I was able to measure both influx and extrusion of Ca during single contraction cycles and relate Ca extrusion to late depolarizations in the action potential. From several suggested mechanisms, only electrogenic Na/Ca exchange could account for the data. This work was completed and published (Hilgemann 1986b) almost simultaneously with the first measurements of exchanger currents by Junko Kimura, Akinori Noma and Hiroshi Irisawa. (Kimura *et al.,* 1986)

He could therefore see that, in the DiFrancesco–Noble model, early extrusion is exactly what is predicted since the sodium–calcium exchange current is activated as soon as the calcium signal occurs, which is responsible also for activating the contractile machinery. It is obvious when one thinks about it. The calcium system is constructed to activate the contractile proteins as quickly as possible. There is no reason why that should not also activate the sodium–calcium exchange proteins.

Don's proposal was direct and to the point. He wanted to come to Oxford as soon as possible so that we could work together to follow up on what was obviously a novel and important insight. He had the experiments. We had the explanation. Don came, and so my life became a fine balance between maintaining an active, and sometimes, controversial laboratory while also carrying the even more controversial national media role on behalf of science and universities in the UK. The political side of what I was doing had also become important. After a year or so of the standoff, the government eventually took me and my colleagues in Save British Science seriously. Not only did I interact effectively with the Science Minister (then Robert Jackson) but I also interacted with the Prime Minister's personal science advisor, George Guise. The stakes were high. Meanwhile, Julian Jack (see Chapter 3) became a Trustee of the Wellcome Trust, which transformed itself into the giant funding charity that it is today, and which also greatly influenced later government policy. In fact, those initiatives eventually led to the sustained recovery of science funding that occurred during the Blair governments, particularly while

David Sainsbury was the Minister of Science. So, maybe it was 'wise' after all. But it took all of two decades to find that out.

5.11. The Hilgemann–Noble Model

In between the politics, Don and I set to work. The first problem was that he was not working on Purkinje tissue. He was working on the rabbit atrium. So, we decided to develop the DiFrancesco–Noble model to be applicable to the very different electrical waveforms found in the atrium. Don is a man of many parts. In addition to his consummate experimental ability (he went on to later invent the giant patch technique (Collins *et al.*, 1992; Hilgemann, 1989, 1990)), he has great computational skill. I convinced IBM to donate an early version of its PC to us since my lab was then too poor to buy one – we had also suffered from the funding cutbacks. Don bought his own PC and wrote the programs in TurboPascal, which eventually convinced me to switch from Algol, which I had been using since the development of more structured programming languages. The days of writing machine code were over, for me, at least.

Painfully slowly (the graphs were just dots on the screen that appeared one after the other after the long integration steps), the model was adjusted. This was the first time we had used a PC. Up till that time, all the computational work had been done on big mainframe computers. Often enough, the early PC (no Windows, just MS-DOS) was left to work through the night. I also used it to develop the first publicly distributed program for heart modelling, OXSOFT HEART. I had formed a company in 1984 to market it. I would have preferred to distribute it for free, but storage discs in those days were far too expensive at around £100 each.

It didn't take long for Don to develop an atrial model from the Purkinje one and to confirm the fit to his experimental findings (Fig. 5.7). But, fortunately for the future work on cardiac cell modelling, he did not stop there. He moved the boundaries of the model deep into further protein systems in the cardiac cell. First, we incorporated the calcium buffer proteins so that the total quantities of calcium being cycled were closer to reality. Then, we completely overhauled the sarcoplasmic reticulum system using what we could of the experimental results of Fabiato (1983), who had performed some heroic experiments in which he removed the cell membrane so that he could study directly the process (calcium-induced calcium release) by which the full calcium signal is formed.

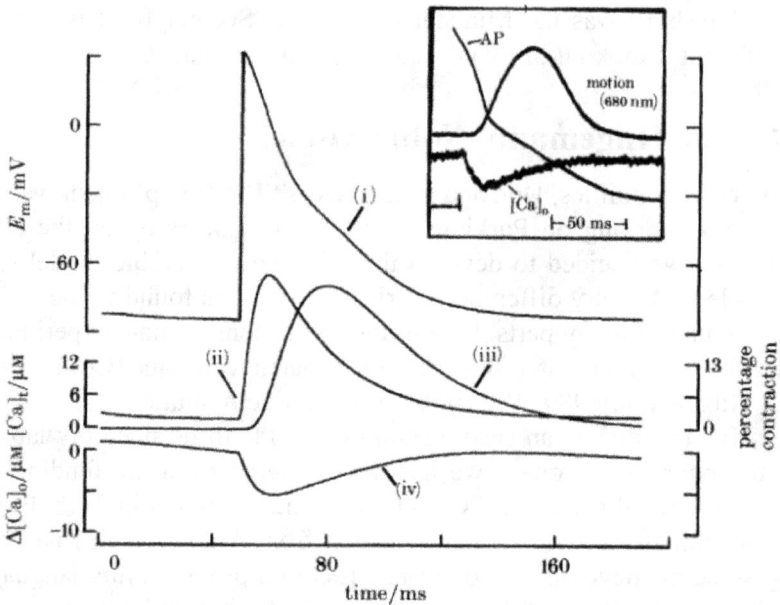

Figure 5.7.   The Hilgemann–Noble 1987 model of the (i) atrial action potential (AP), showing its reconstruction of the (ii) intracellular calcium transient $[Ca]_i$, (iii) contraction (Motion) and (iv) the extracellular calcium transient $[Ca]_o$. The inset shows one of Hilgemann's experimental recordings showing how close the correspondence between the experiment and the model is for the extracellular calcium transient.

5.12. Suppers and Wine at Holywell Manor[5]

Much of this work, and that with Dario DiFrancesco, was carried out over impromptu suppers and convivial wines at Holywell Manor in Oxford, where Balliol College has its Graduate Centre. At the time, I was the head

---

[5]My home and kitchen have been the focus of such lab conviviality ever since the days of Dick Tsien, who introduced Chinese cooking. The rolls of the i_{Kr} and i_{Ks} recordings described in Chapter 2 were brought directly from the laboratory to be laid out on the floor of my home. Julian Jack, during the writing of *Electric Current Flow in Excitable Cells*, introduced us to the joys of Barsac served with a kiwi fruit pavlova. Otto Hauswirth cooked Szegedina Goulash, sent as a recipe from his mother in Vienna. Carlos Ojeda produced amazing couscous, while I cooked the endless Indian curries.

(Praefectus) of the Centre, living in its beautiful lodgings, a manor house dating from 1516. At the completion of our work and its submission to the journal, Don and I celebrated, not only with a good bottle of Bordeaux but with what I believe was a cognac with a vintage around the time that the Allies defeated Hitler in the Second World War. I must have saved the bottle for some occasion like this one.

In retrospect, we had every reason to celebrate. Those improvements in the modelling have turned out to be just as seminal as the DiFrancesco–Noble model. In fact, I see the different models, all the way from 1960 to 1990, as forming a continuum of development, sometimes occurring in a revolutionary way, but often also as a slow evolution of detail in the continuous interaction between experiment and theory. The experimental work in my lab, particularly on sodium–calcium exchange (Ch'en *et al.*, 1998; Chapman and Noble, 1989; Egan *et al.*, 1989; Milberg *et al.*, 2008; Noble, 1996, 2002a, 2002b; Noble and Blaustein, 2007; Noble *et al.*, 1991, 1996, 2007; Noble and Powell, 1991; Noble and Varghese, 1998; Sher *et al.*, 2007), continued to benefit from this interaction over all the subsequent years.

5.13. The First Single-Cell Model

All the models described so far were developed from experiments performed on intact tissue. By the time Don Hilgemann was working with us, Trevor Powell had come to Oxford as a British Heart Foundation Professor. He had pioneered the technique of isolating ventricular cardiac cells so that experiments could be performed on single cells using microelectrodes or micro patch electrodes. We published a book together on the exploitation of his discovery (Noble and Powell, 1987). The problems associated with multicellular preparations then became a thing of the past, and the powerful technique of patch clamping allowed recordings from single cells to be made either in whole-cell mode or from single channels within the patch itself.

Parallel to the isolation of ventricular cells, the technique of obtaining single pacemaker cells from the SA node was developed, and our group made progress using these as well (Denyer and Brown, 1990a, 1990b). Quite quickly, all the work in our laboratory and around the world shifted

Figure 5.8.   Susan Noble, Junko Kimura, Denis Noble and Yung Earm (left to right) at a meeting in Japan in 1985.

to the single-cell techniques. This created a need for the models to also apply to single cells. One of my collaborators at that time was Yung Earm (Fig. 5.8) from Seoul National University in Korea. He had previously worked with me between 1979 and 1981, so he already knew the Oxford lab well. His second visit was on a British Heart Foundation Visiting Fellowship, and it was during this time that we worked on developing the first single-cell model (Earm and Noble, 1990).

He and his colleagues (Earm *et al.,* 1990) subsequently used this in a lovely experiment that would shed more light on the role of sodium–calcium exchange. They performed an experiment in which they allowed the early phases of the atrial action potential to run as usual. But then, at the beginning of the low plateau phase, they clamped the voltage at a level (−40 mV) that enables the slow development and decay of the inward exchange current to be recorded. They then infused a calcium chelator into the cell through the electrode to show that, once the calcium transient was removed, the current was as well. A single good experiment confirmed the essentials of what the model demonstrated (Fig. 5.9).

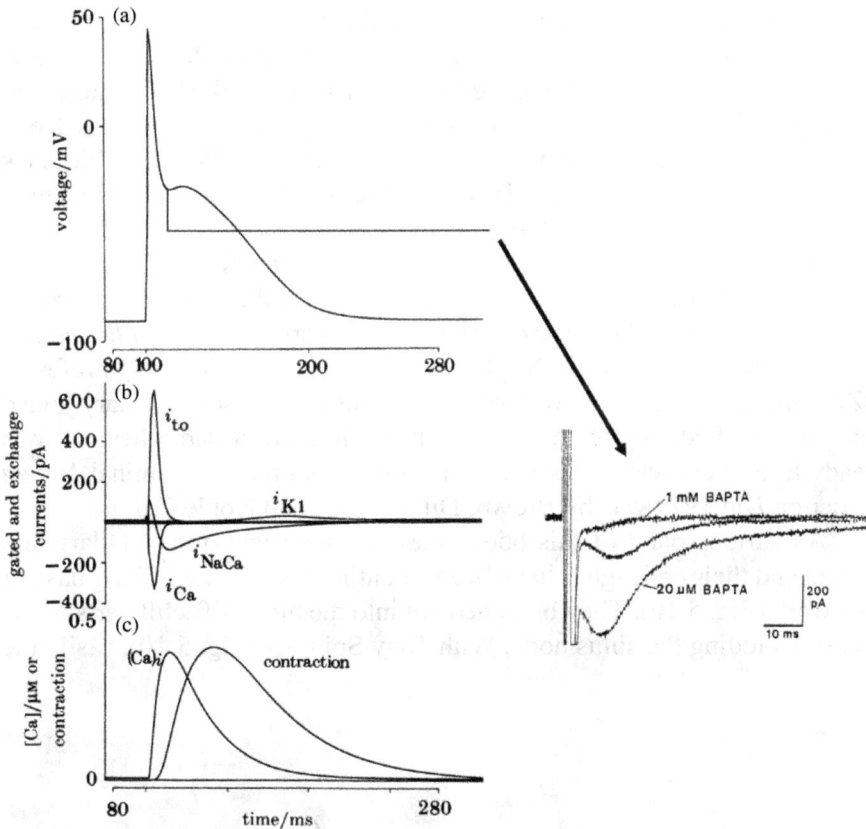

Figure 5.9. The single atrial cell model (Earm and Noble, 1990) and the experimental results (Earm *et al.*, 1990) showing the time course of the sodium–calcium exchange current during a voltage clamp at the time of the late low plateau phase of the action potential. (a) voltage changes; (b) ionic current changes; (c) intracellular calcium and contraction changes.

5.14. Women in Physiology

It will be obvious even to a casual reader that, in the work described in this central chapter, women physiologists played a major role. In the earliest years of my Oxford laboratory, women (both students and faculty) were confined to the 'women's colleges'. Somerville College had the good fortune to have Jean Banister as the medical and physiology fellow and tutor. Her main field was cardiovascular, so it was natural to link up with her. We famously did live demonstrations together for the undergraduate students. Jean performed the experiment while I tried to time the pivotal lecture points to arrive just in time for the experiment to illustrate it. Most of the time, the synchronisation worked, but there was much hilarity amongst the students when we got it wrong, so that I was convincingly announcing that x had happened, while Jean hit back sharply, 'No, it's y!'. The tortoise heart experiment illustrated in Chapter 4 (Fig. 4.1) was from one of those demonstrations in 1969.

She was tutor to two of the scientists who worked in my group during these seminal years. One was my wife, Susan. The other was Hilary Brown, who had worked on her thesis on the heart of a crustacean, *Squilla mantis*, in the marine laboratory in Naples, where my former Anatomy Professor JZ Young did most of his work on the octopus and the squid. Hilary joined up with my fledgling team and was then joined by Susan when she was ready to do her thesis. This is how it came about that the seminal *Nature* paper on i_f in 1979 was by Brown, DiFrancesco and Noble (Susan).

As various parts of this book show, the contributions of Hilary and Susan and their colleagues have been a continuous and flourishing base of the work (Fig. 5.10). They branched out into the more difficult parts of the heart, including the sinus node. With Tony Spindler (Fig. 5.10), easily the

Figure 5.10. Hilary Brown (left), Junko Kimura (middle front), David Eisner (middle back) and Anthony Spindler (right).

longest-serving member of the team and the mainstay of all our developments of electrophysiological technology, they developed our lab's version of a technique known as the sucrose gap for doing voltage clamp experiments on tissues where microelectrode work was too difficult. Susan taught this technique to Wayne Giles, one of the authors of the *Nature* paper on the T wave described in Chapter 4. Wayne and Susan worked extremely well together, and their collaboration was crowned with a key paper showing the inhibitory influence of acetylcholine on the calcium current in the heart (Giles and Noble, 1976).

Tony Spindler was renowned for developing one of the best sucrose gap rigs in the world. The Perspex halves of the compartments were machined and polished to produce optically flat surfaces so that, when they were brought together, a tight fit ensured a very high resistance, forcing the electric current to canalise itself through the cells of the tissue rather than leak through the compartments. People from our lab took these beautiful pieces of apparatus to their own labs around the world when they left. Although Tony originally joined me as a technician, he proudly converted himself into a fully-fledged scientist and gained his doctorate in 2000, a year or two before he retired. The focus of his thesis was on i_{bNa} whose role in pacemaker activity is described in Chapter 6. *Doctor* Spindler, indeed (Spindler *et al.,* 1998).

In addition to the science, Hilary and Susan were also responsible, directly and indirectly, for a statistic we are all proud of, which is that during almost the whole period, the proportion of women scientists working in the group became and remained substantial. This was not achieved as any kind of campaign. It happened naturally. Their presence as leading researchers in the team encouraged others to come.

There are two very appropriate endings to this part of the story. The first is that, as the newly elected President of IUPS at its Kyoto Congress in 2009, I attended a symposium on Women in Physiology organised by our former colleague, Junko Kimura (who also had Jean Banister as her tutor at Somerville College), whose work on sodium–calcium exchange has already been described in this chapter. That meeting produced a published report (Kimura *et al.*, 2010) to be considered by the Council of the Union. It highlights the wide variations in gender inequality in various parts of the world. Japan itself has one of the worst statistics. Junko herself was rare as a professor and head of department.

The second is that one of the last of the women to work in the group is my own daughter, Penny, who has been a mainstay of the programming and curating work all the way from taking over the developments of OXSOFT HEART to curating the models on the CellML website and working on their use in problems of cardiac repolarisation (Garny *et al.*, 2002; Liu *et al.*, 2008; Noble and Noble, 2006; Noble *et al.*, 2007, 2010; Noble and Noble, 2000, 2010; Roux *et al.*, 2001, 2006; Sears *et al.*, 1999, 2007, 2008; Ten Tusscher *et al.*, 2004; Volk *et al.*, 2005). Her work in the laboratory ceased when my active laboratory closed for the final time just before the pandemic.

5.15. Conclusions

5.15.1. *Pluses*

This chapter forms the central core of the work, with a flowering of inter-action between experiment and theory that generated the whole family of cardiac cell models developed since that period. Given the controversy that occurred both in the development of the work and subsequently, it is remarkable how much of it has completely stood the test of time. Without this work, the Cardiac Physiome Project (Chapter 7) could not have developed as it has and would not have done so on a firm experimental and modelling basis at the cellular level. Of course, this flowering of the field depended on much more than the Oxford laboratory.

5.15.2. *Minuses*

We could have achieved much more if funding had allowed it. Streams of good people were clamouring to work in our labs. In fact, without the secure and unwavering support of the British Heart Foundation, which gave me a chair and associated funds from 1984 to 2004, the work would have collapsed. Government funding alone would have been inadequate. I shall always be grateful for the BHF's support. That Foundation is the reason why cardiac science still flourishes in the UK, while the Wellcome Trust did so for medical science as a whole.

5.15.3. *Contributions to systems biology*

There is much argument today about how systems biology should be conducted. The work described in this chapter shows how rich the rewards can be when experimental work and theoretical work interact in close proximity, with some people doing both. As this story illustrates, at some stages, theory was ahead. At other stages, the experimental work triggered new developments in theory. In Chapter 9, I will elaborate on the middle-out approach to multilevel analysis in biology. This chapter shows that cell modelling clearly forms my middle, from which we were able to subsequently reach out to both lower and higher levels in what I believe to be one of the major paradigm examples of systems biology in practice.

The work of this period also illustrates the role of analytical mathematics in complementing brute-force heavy computing. The one regret I have is that, since so few of my physiological colleagues understand what we were doing the crucial role of maths has easily been forgotten.

References

Arlock, P. and Noble, D. (1985) 'Two components of "second inward current" in ferret papillary muscle', *Journal of Physiology*, 369, p. 88.

Beeler, G. W. and Reuter, H. (1977) 'Reconstruction of the action potential of ventricular myocardial fibres', *Journal of Physiology*, 268, pp. 177–210.

Brown, H. F., Clark, A. and Noble, S. J. (1976a) 'Analysis of pacemaker and repolarization currents in frog atrial muscle', *Journal of Physiology*, 258, pp. 547–577.

Brown, H. F., Clark, A. and Noble, S. J. (1976b) 'Identification of the pacemaker current in frog atrium', *Journal of Physiology*, 258, pp. 521–545.

Brown, H. F. and DiFrancesco, D. (1980) 'Voltage-clamp investigations of membrane currents underlying pace-maker activity in rabbit sinoatrial node', *Journal of Physiology*, 308, pp. 331–351.

Brown, H. F., DiFrancesco, D. and Noble, S. J. (1979) 'How does adrenaline accelerate the heart?', *Nature*, 280, pp. 235–236.

Brown, H. F., Kimura, J., Noble, D., Noble, S. J. and Taupignon, A. (1983) 'Two components of "second inward current" in the rabbit S.A. node', *Journal of Physiology*, 334, pp. 56–57.

Brown, H. F., Kimura, J., Noble, D., Noble, S. J. and Taupignon, A. (1984a) 'The ionic currents underlying pacemaker activity in rabbit sino-atrial node: Experimental results and computer simulations', *Proceedings of the Royal Society B*, 222, pp. 329–347.

Brown, H. F., Kimura, J., Noble, D., Noble, S. J. and Taupignon, A. (1984b) 'Mechanisms underlying the slow inward current, isi, in the rabbit sino-atrial node investigated by voltage clamp and computer simulation', *Proceedings of the Royal Society B*, 222, pp. 305–328.

Brown, H. F., Noble, D., Noble, S. J. and Taupignon, A. (1984c) 'Transient inward current and its relation to the very slow inward current in the rabbit S.A. node', *Journal of Physiology*, 349, pp. 47.

Ch'en, F. C., Vaughan-Jones, R. D., Clarke, K. and Noble, D. (1998) 'Modelling myocardial ischaemia and reperfusion', *Progress in Biophysics and Molecular Biology*, 69, pp. 515–537.

Chapman, R. A. and Noble, D. (1989) 'Sodium-Calcium exchange in the heart', In *Sodium-Calcium Exchange* (ed. J. Allen, D. Noble and H. Reuter), pp. 102–125. Oxford: Oxford University Press.

Christoffels, V. M., Smits, G. J., Kispert, A. and Moorman, A. F. M. (2010) 'Development of pacemaker tissues of the heart', *Circulation Research*, 106, pp. 240–254.

Collins, A., Somlyo, A. V. and Hilgemann, D. W. (1992) 'The giant cardiac membrane patch method: Stimulation of outward $Na^{(+)}$-Ca^{2+} exchange current by MgATP', *Journal of Physiology*, 454, pp. 27–57.

Denyer, J. C. and Brown, H. F. (1990a) 'Pacemaking in rabbit isolated sion-atrial node cells during Cs^+ block of the hyperpolarization-activated current if', *Journal of Physiology*, 429, pp. 401–409.

Denyer, J. C. and Brown, H. F. (1990b) 'Rabbit sino-atrial node cells: Isolation and electrophysiological properties', *Journal of Physiology*, 428, pp. 405–424.

DiFrancesco, D. (2010) 'The role of the funny current in pacemaker activity', *Circulation Research*, 106, pp. 434–446.

DiFrancesco, D. and Noble, D. (1980a) 'If i_{K2} is an inward current, how does it display potassium specificity?', *Journal of Physiology*, 305, p. 14.

DiFrancesco, D. and Noble, D. (1980b) 'The time course of potassium current following potassium accumulation in frog atrium: Analytical solutions using a linear approximation', *Journal of Physiology*, 306, pp. 151–173.

DiFrancesco, D. and Noble, D. (1982) 'Implications of the re-interpretation of i_{K2} for the modelling of the electrical activity of pacemaker tissues in the heart', In *Cardiac Rate and Rhythm* (ed. L. N. Bouman and H. J. Jongsma), pp. 93–128. The Hague, Boston, London: Martinus Nijhoff.

DiFrancesco, D. and Noble, D. (1985) 'A model of cardiac electrical activity incorporating ionic pumps and concentration changes', *Philosophical Transactions of the Royal Society B*, 307, pp. 353–398.

Earm, Y. E., Ho, W. K. and So, I. S. (1990) 'Inward current generated by Na-Ca exchange during the action potential in single atrial cells of the rabbit', *Proceedings of the Royal Society B*, 240, pp. 61–81.

Earm, Y. E. and Noble, D. (1990) 'A model of the single atrial cell: Relation between calcium current and calcium release', *Proceedings of the Royal Society B*, 240, pp. 83–96.

Efimov, I. R., Federov, V. V., Joung, B. and Lin, S.-F. (2010) 'Mapping cardiac pacemaker circuits: Methodological puzzles of the sino-atrial node optical mapping', *Circulation Research*, 106, pp. 255–271.

Egan, T., Noble, D., Noble, S. J., Powell, T., Spindler, A. J. and Twist, V. W. (1989) 'Sodium-Calcium exchange during the action potential in guinea-pig ventricular cells', *Journal of Physiology*, 411, pp. 639–661.

Eisner, D. A., Choi, H. S., Diaz, M. E. and O'Neill, S. C. (2000) 'Integrative analysis of calcium cycling in cardiac muscle', *Circulation Research*, 87, pp. 1087–1094.

Eisner, D. and Sipido, K. R. (2004) 'Sodium calcium exchange in the heart. Necessity or luxury?', *Circulation Research*, 95, pp. 549–551.

Fabiato, A. (1983) 'Calcium induced release of calcium from the sarcoplasmic reticulum', *American Journal of Physiology*, 245, pp. C1–14.

Fedida, D., Noble, D., Shimoni, Y. and Spindler, A. J. (1987) 'Inward currents related to contraction in Guinea-pig ventricular myocytes', *Journal of Physiology*, 385, pp. 565–589.

Garny, A., Noble, P. J., Kohl, P. and Noble, D. (2002) 'Comparative study of sino-atrial node cell models', *Chaos, Solitons and Fractals*, 13, pp. 1623–1630.

Giles, W. and Noble, S. J. (1976) 'Changes in membrane currents in bull-frog atrium produced by acetylcholine', *Journal of Physiology*, 261, pp. 103–123.

Hilgemann, D. W. (1986a) 'Extracellular calcium transients and action potential configuration changes related to post-stimulatory potentiation in rabbit atrium', *Journal of General Physiology*, 87, pp. 675–706.

Hilgemann, D. W. (1986b) 'Extracellular calcium transients at single excitations in rabbit atrium measured with tetramethylmurexide', *Journal of General Physiology*, 87, pp. 707–735.

Hilgemann, D. W. (1989) 'Giant excised cardiac sarcolemmal membrane patches: Sodium and sodium-calcium exchange currents', *Pflügers Archiv, European Journal of Physiology*, 415, pp. 247–249.

Hilgemann, D. W. (1990) 'Regulation and deregulation of cardiac Na-Ca exchange in giant excised sarcolemmal membrane patches', *Nature*, 344, pp. 242–245.

Hilgemann, D. W. and Noble, D. (1987) 'Excitation-contraction coupling and extracellular calcium transients in rabbit atrium: Reconstruction of basic cellular mechanisms', *Proceedings of the Royal Society B*, 230, pp. 163–205.

Johnson, E. A. and Lieberman, M. (1971) 'Heart: Excitation and contraction', *Annual Reviews of Physiology*, 33, pp. 479–530.

Kimura, J., Miyamae, S. and Noma, A. (1987) 'Identification of sodium-calcium exchange current in single ventricular cells of guinea-pig', *Journal of Physiology*, 384, pp. 199–222.

Kimura, J., Noma, A. and Irisawa, H. (1986) 'Na-Ca exchange current in mammalian heart cells', *Nature*, 319, pp. 596–597.

Kimura, J., Suzuki, Y., Mizumura, K., Katagiri, C., Azuma, K., Hao, L., Gulia, K., Wray, S., Iguchi-Ariga, S. and Barrett, K. (2010) 'Report on the Women in Physiology Symposium in IUPS 2009', *Journal of Physiological Sciences*, 60, pp. 227–234.

Lakatta, E. G., Maltsev, V. A. and Vinogradova, T. M. (2010) 'A coupled system of intracellular Ca2+ clocks and surface membrane ion clocks controls the timekeeping of the heart's pacemaker', *Circulation Research*, 106, pp. 659–673.

Lederer, W. J. and Tsien, R. W. (1976) 'Transient inward current underlying arrhythmogenic effects of cardiotonic steroids in Purkinje fibres', *Journal of Physiology*, 263, pp. 73–100.

Liu, J., Noble, P. J., Xiao, G., Mohamed, A., Dobrzynski, H., Boyett, M. R., Lei, M. and Noble, D. (2008) 'Role of pacemaking current in cardiac nodes: Insights from a comparative study of sinoatrial node and atrioventricular node', *Progress in Biophysics and Molecular Biology*, 96, pp. 294–304.

McAllister, R. E., Noble, D. and Tsien, R. W. (1975) 'Reconstruction of the electrical activity of cardiac Purkinje fibres', *Journal of Physiology*, 251, pp. 1–59.

Milberg, P., Pott, C., Fink, M., Frommeyer, G., Matsuda, T., Baba, A., Osada, N., Breithardt, G., Noble, D. and Eckardt, L. (2008) 'Inhibition of the Na^+/Ca^{2+} exchanger suppresses torsade de pointes in an intact heart model of LQT2 and LQT3', *Heart Rhythm*, 5, pp. 1444–1452.

Mullins, L. (1981) *Ion Transport in the Heart*. New York: Raven Press.

Noble, D. (1984) 'The surprising heart: A review of recent progress in cardiac electrophysiology', *Journal of Physiology*, 353, pp. 1–50.

Noble, D. (1996) 'The functional significance of sodium-calcium exchange in the heart', In *Molecular Physiology and Pharmacology of Cardiac Ion Channels and Transporters* (ed. M. Morad, S. Ebashi, W. Trautwein and Y. Kurachi), pp. 457–467: Kluwer.

Noble, D. (2002a) 'Influence of Na-Ca exchange stoichiometry on model cardiac action potentials', *Annals of the New York Academy of Sciences*, 976, pp. 133–136.

Noble, D. (2002b) 'Simulation of Na-Ca exchange activity during ischaemia', *Annals of the New York Academy of Sciences*, 976, pp. 431–437.

Noble, D. and Blaustein, M. P. (2007) 'Directionality in drug action on sodium-calcium exchange', *Annals of the New York Academy of Sciences*, 1099, pp. 540–543.

Noble, D., LeGuennec, J. Y. and Winslow, R. (1996) 'Functional roles of sodium-calcium exchange in normal and abnormal cardiac rhythm', *Annals of the New York Academy of Sciences*, 779, pp. 480–488.

Noble, D. and Noble, P. J. (2006) 'Late sodium current in the pathophysiology of cardiovascular disease: Consequences of sodium-calcium overload', *Heart*, 92, pp. iv1–iv5.

Noble, D., Noble, P. J. and Fink, M. (2010) 'Competing oscillators in cardiac pacemaking: Historical background', *Circulation Research*, 106, pp. 1791–1797.

Noble, D. and Noble, S. J. (1984) 'A model of S.A. node electrical activity using a modification of the DiFrancesco-Noble (1984) equations', *Proceedings of the Royal Society B*, 222, pp. 295–304.

Noble, D., Noble, S. J., Bett, G. C. L., Earm, Y. E., Ho, W. K. and So, I. S. (1991) 'The role of sodium-calcium exchange during the cardiac action potential', *Annals of the New York Academy of Sciences*, 639, pp. 334–353.

Noble, D. and Powell, T. (1987) *Electrophysiology of Single Cardiac Cells*. London: Academic Press.

Noble, D. and Powell, T. (1991) 'The slowing of calcium signals by calcium indicators in cardiac muscle', *Proceedings of the Royal Society B*, 246, pp. 167–172.

Noble, D., Sarai, N., Noble, P. J., Kobayashi, T., Matsuoka, S. and Noma, A. (2007) 'Resistance of Cardiac Cells to NCX Knockout: A Model Study', *Annals of the New York Academy of Sciences*, 1099, pp. 306–309.

Noble, D. and Varghese, A. (1998) 'Modeling of sodium-calcium overload arrhythmias and their suppression', *Canadian Journal of Cardiology*, 14, pp. 97–100.

Noble, P. J. and Noble, D. (2000) 'Reconstruction of the cellular mechanisms of cardiac arrhythmias triggered by early after-depolarizations', *Japanese Journal of Electrocardiology*, 20 (Suppl 3), pp. 15–19.

Noble, P. J. and Noble, D. (2010) 'A Historical Perspective on the Development of Models of Rhythm in the Heart', In *Heart Rate and Rhythm: Molecular Basis, Pharmacological Modulation and Clinical Implications* (ed. O. N. Tripathi, U. Ravens and M. C. Sanguinetti), pp. (27–42). Heidelberg: Springer.

Noble, S. J. (1976) 'Potassium accumulation and depletion in frog atrial muscle', *Journal of Physiology*, 258, pp. 579–613.

Noma, A. and Irisawa, H. (1976) 'Membrane currents in the rabbit sinoatrial node cell as studied by the double microelectrode method', *Pflügers Archiv, European Journal of Physiology*, 364, pp. 45–52.

O'Rourke, B. (2010) 'Be still my beating heart: Never!', *Circulation Research*, 106, pp. 238–239.

Provencher, S. W. (1976) 'An eigenfunction expansion method for the analysis of exponential decay curves', *Journal of Chemical Physics*, 64, p. 2772.

Reuter, H. (1967) 'The dependence of slow inward current in Purkinje fibres on the extracellular calcium concentration', *Journal of Physiology*, 192, pp. 479–492.

Reuter, H. and Seitz, N. (1969) 'The dependence of calcium efflux from cardiac muscle on temperature and external ion composition', *Journal of Physiology*, 195, pp. 451–470.

Roux, E., Noble, P. J., Hyvelin, J.-M. and Noble, D. (2001) 'Modelling of Ca^{2+}-activated chloride current in tracheal smooth muscle cells', *Acta Biotheoretica*, 49, pp. 291–300.

Roux, E., Noble, P. J., Noble, D. and Marhl, M. (2006) 'Modelling of calcium handling in airway myocytes', *Progress in Biophysics and Molecular Biology*, 90, pp. 64–87.

Sears, C. E., Noble, D., Noble, P. J. and Paterson, D. (1999) 'Vagal control of heart rate is modulated by extracellular potassium', *Journal of the Autonomic Nervous System*, 77, pp. 164–171.

Seyama, I. (1976) 'Characteristics of the rectifying properties of the sino-atrial node cell of the rabbit', *Journal of Physiology*, 255, pp. 379–397.

Sher, A., Hinch, R., Noble, P. J., Gavaghan, D. and Noble, D. (2007) 'Functional significance of Na^+/Ca^{2+} exchangers co-localisation with ryanodine receptors', *Annals of the New York Academy of Sciences*, 1099, pp. 215–220.

Shrier, A. and Clay, J. R. (1982) 'Comparison of the pacemaker properties of chick embryonic atrial and ventricular heart cells', *Journal of Membrane Biology*, 69, pp. 49–56.

Spindler, A. J., Noble, S. J., Noble, D. and LeGuennec, J. Y. (1998) 'The effects of sodium substitution on currents determining the resting potential in guinea-pig ventricular cells', *Experimental Physiology*, 83, pp. 121–136.

Ten Tusscher, K. H. W. J., Noble, D., Noble, P. J. and Panfilov, A. V. (2004) 'A model of the human ventricular myocyte', *American Journal of Physiology*, 286, pp. 1573–1589.

Venetucci, L. A., Trafford, A. W. and Eisner, D. A. (2007) 'Increasing ryanodine receptor open probability alone does not produce arrhythmogenic calcium waves: Threshold sarcoplasmic reticulum calcium content is required', *Circulation Research*, 100, pp. 105–111.

Volk, T., Noble, P. J., Wagner, M., Noble, D. and Ehrmke, H. (2005) 'Ascending aortic stenosis selectively increases action potential-induced Ca^{2+} influx in epicardial myocytes of the rat left ventricle', *Experimental Physiology*, 90, pp. 111–121.

Chapter 6

Understanding Robustness in Biological Systems

6.1. Background: The Genome 10 Years On

When I first drafted this chapter, the June 2010 edition of *Prospect* carried an article on the genome headed 'Too much information'. It begins as follows:

> Ten years ago, the first draft of the sequence of the human genome was heralded as the dawn of a new era of genetic medicine.... You might have noticed that it hasn't [happened]. The medical impact of the human genome project (HGP) has so far been negligible.

The explanation given by the author, no doubt following a lead from an editorial in *Nature* (Editorial, 2010), is thus:

> The activity of genes is affected by many things not explicitly encoded in the genome, such as how the chromosomal material is packaged up and how it is labelled with chemical markers. Even for diseases like diabetes, which have a clear inherited component, the known genes involved seem to account for only a small proportion of the inheritance... the failure to anticipate such complexity in the genome must be blamed partly on the cosy fallacies of genetic research. After Francis Crick and James Watson cracked the riddle of DNA's molecular structure in 1953, geneticists could not resist assuming it was all over bar the shouting. They began to see DNA as the "book of life," which could be read like

an instruction manual. It now seems that the genome might be less like a list of parts and more like the weather system, full of complicated feedbacks and interdependencies.

In 2002, I wrote an article for the magazine of The Physiological Society, *Physiology News,* explaining why the genome is not the 'Book of Life' (Noble, 2002). A reader was so enthused by it that he approached the editor of *Prospect* to insist that it deserved a much wider readership and thought that *Prospect* was the ideal medium. It would have been, but the offer was turned down without question. It didn't fit the mindset of that time, full of confidence that molecular biology was going, finally, to deliver the goods through the exploitation of the genome data.

What had changed in the subsequent 8 years, so that even a staff writer for *Prospect* now expresses the main message of the 2002 article? The answer is that 2010 is the 10-year watermark after the sequencing of the genome, when we were promised by the leaders of the Human Genome Project that the benefits for health care would have arrived. Diabetes, hypertension and mental illness were amongst the targets. Science journalists are therefore becoming uneasy that they bought into a promise that has not and, I would argue, could not have been delivered. And they are not alone. The drug industry also bought in, literally so, since start-up genomics companies were bought for hundreds of millions of dollars. The sequencing of genomes has been of great value for basic science, particularly in studies on the comparison of genomes for the purposes of evolutionary biology, but the interpretation of the genome data in terms of biological functions (phenotypes) has proved vastly more difficult than anticipated.

As I revise this chapter in 2025, it has become even clearer that we backed the wrong horse in the battle to find cures for cancer, cardiovascular diseases, nervous diseases and much else. The top science journal, *Nature,* invited me in early 2024 to review a book (Ball, 2024) detailing, in over 500 pages, the many phenomenal discoveries in biology that simply do not fit the gene-centric biology of the 20th century. My article was published under the headline 'Genes are not the Blueprint for Life' (Noble, 2024). Just a year earlier, in 2023, my brother, Raymond Noble, and I published a short popular book by Cambridge University Press, called *Understanding Living Systems,* explaining the same message in easy-to-read non-technical language (Noble and Noble, 2023). Just after that book was published, a team at University College London published an article describing a critical test: do the polygenic scores succeed in predicting disease states? (Hingorani *et al.*, 2023). These are obtained by summing together all the association scores between the presence and absence of particular gene variants with a specific disease state in later life. The answer was crystal clear. Yes, there were successful predictions of cancer or cardiovascular disease, but there were also far too many false positives, where the prediction turned out to be false. By the same standards as are used for assessing the efficacy of a new drug by the FDA, this was a failure. These so-called predictions are causing anxiety in people that is simply not justified.

6.2. Limitations of the Differential View of Genetics

Physiologists have been aware of many of the relevant forms of 'complicated feedbacks and interdependencies' for many years. We were largely ignored as molecular biology took centre stage during the second half of the 20th century. One of the reasons is an unnecessary and incorrect limitation on the relationship between genes and phenotypes that developed in 20th century biological thought. The standard story in the neo-Darwinism view of evolution is that differences in genes (different alleles) are responsible for differences in phenotype, so that selection of the phenotype equates with selection of the successful allele. Evolution is seen as occurring primarily through incremental change as new alleles (mutations) arise. I call this the *differential* view of genetics (Noble, 2011a). It is the basis of the selfish gene concept (Dawkins, 1976, 2006). In a later book, *The Extended Phenotype*, however, Dawkins acknowledges that no experiment could distinguish between the gene-centred view of evolution and its alternatives (Dawkins, 1982, page 1).

The differential view of genetics is limiting and somewhat analogous to where we would be in mathematics if we were to be limited to differential equations and if the integral sign had never been invented. The analogy is quite good. To integrate differential equations, we need the initial and boundary conditions. These are just as much a 'cause' of the particular solutions we obtain as the differential equations themselves. Likewise, differences in genomes operate in the context of the system as a whole, which form the initial (egg cell) and boundary (environmental) conditions for the development of an organism. Those conditions are just as much a cause of the phenotype as are the genes (Noble, 2008; Kohl *et al.*, 2010).

Moreover, nature does not recognise the limitation of the differential view. A gene, defined as a particular sequence of DNA, has many effects. These are not limited to the differences one can observe between that gene and one of its alleles. For a full physiological understanding of the relationship between genomes and phenotypes, we therefore need an *integral* view of genetics (Noble, 2011a), in which we look for the many phenotypic effects it may have, even when those phenotypic effects are hidden by the system.

When I first read Richard Dawkins's acknowledgement in *The Extended Phenotype* ('I doubt that there is any experiment that could be done to prove my claim'), I was strongly inclined to agree with it and, indeed, if you compare the selfish gene metaphor with an opposing metaphor, such as genes as prisoners, it is impossible to think of an experiment that would distinguish between the two views, as I show in my book *The Music of Life* (Noble, 2006, Chapter 1). For any given case, I still think that must be true. But I have slowly changed my view on whether this must be true if we consider *many* cases, looking at the functioning of the organism as a whole. There are different ways in which empirical discovery can impact our theoretical understanding. Not all of these are in the form of the straight falsification of a hypothesis. Sometimes, it is the slow accumulation of the weight of evidence that eventually triggers a change of viewpoint. This is the case with insights that are expressed in metaphorical form (like 'selfish' and 'prisoners'), not intended to be taken literally.

Consider the following thought experiment. Take an organism for which we know the complete genome. Check on the proteins for which each of the genes forms a template. Then, carry out knockout experiments in turn on every single one of the genes in the organism. Suppose we find that in the great majority of cases, the knockouts have no effect whatsoever! Would we conclude that these were not, after all, genes since changing them did not have a phenotypic effect? Clearly not. Each and every one of them forms templates for the production of proteins that function in particular networks. That is what we now mean by a 'gene'. At the level of proteins therefore all the genes can be expressed and have functionality in that sense. But at the level of the phenotype, they appear not to be functional. I think that the sheer weight of such experimental evidence would force us to reconsider the relationship between genes and phenotypes. We would conclude that there is no sense in talking just about differences between alleles of particular genes as though that were the sole determinant of a phenotype difference. That would be to mistake the tip for the whole iceberg.

6.3. Robustness in Yeast

Actually, this is not just a thought experiment at all. It is a real one! It was carried out on the 6,000 or so genes in the yeast genome (Hillenmeyer *et al.*, 2008). In the experiments, as many as 80% of knockouts revealed no phenotypic effect in the sense of affecting the growth and reproduction of the organism. Nevertheless, the great majority (97%) of knockouts do have an effect under metabolic stress, when the backup mechanisms are compromised. What is happening here? Modern geneticists call this genetic buffering. The networks of interactions formed by the proteins, the genes and the other structures in the organism with which they all interact are robust. Like aircraft control systems that are fail-safe so that control is not lost when one of the mechanisms fails, organisms are strongly protected against most genetic failures. If one mechanism is knocked out, another one takes over. Moreover, the organism itself continually corrects mistakes in the genome as it gets copied. The robustness of the organism is therefore a property of the system, not of the individual genes, both in terms of protecting the system against failures and repairing copying mistakes.

Now, this fact about the robustness of organisms is clearly an empirical discovery. And equally clearly, it favours the idea of co-operative genes, not selfish genes. Metaphors, for that is what we are dealing with here, may not be falsifiable in the simple way in which an empirical theory is falsifiable. They are nonetheless sensitive to empirical discovery in a more general sense. It is a matter of judgement which metaphor amongst competing metaphors best describes the situation. But that doesn't mean that the judgement is completely arbitrary and independent of the empirical evidence. The relationship is just more nuanced than it would be in a case where a single observation can falsify a scientific hypothesis. This is the reason why, at the end of Chapter 1 of *The Music of Life,* after comparing selfish and cooperative gene metaphors, I wrote, 'it does seem more natural, and certainly more meaningful, to say that the rationale for existence lies at the level at which selection occurs. This is at the level at which we can say why an organism survived or not' (Noble, 2006, page 22).

With this background of ideas in mind, we are ready to consider a particular example and to see what conclusions we can draw from it.

6.4. The Robustness of Heart Rhythm

The article associated with this chapter (Noble *et al.*, 1992) shows how quantitative physiological analysis can reveal the mechanisms of these forms of robustness, how we can reverse engineer a system to avoid the problems raised by the differential view, and therefore precisely why the integrative view of genetics should replace the differential view. Let's analyse this case in detail.

The paper was based on using a model of a sinus node cell derived from the work described in Chapter 5 (Noble and Noble, 1984; DiFrancesco and Noble, 1985). First, we can use the model to answer what seems at first sight to be a simple question. Is there a genetic program for cardiac rhythm? We can answer that question since all the protein mechanisms represented in the equations of the model are made from templates formed by genes. But, try as we may, there is nothing in the DNA sequences for those proteins, nor in the properties of those proteins in isolation, that could form a 'program' for cardiac rhythm.

A simple experiment on the model will demonstrate this.

In Fig. 6.1, the model was run for 1300 ms, during which time 6 oscillations were generated. These correspond to six heartbeats at a frequency similar to that of the heart of a rabbit, the species on which the experimental data were obtained to construct the model. During each beat, all the protein mechanisms also oscillate in a specific sequence. To simplify the diagram, only three of those protein channels are represented here. At 1300 ms, an experiment was performed on the model. The 'downward causation' (see Chapter 1) between the global cell property, the membrane potential and the voltage-dependent gating of the ion channels was interrupted. If there were a subcellular 'program' forcing the proteins to oscillate, the oscillations would continue. In fact, all oscillations cease and the activity of each protein relaxes to a steady value. In this case, therefore, the 'program' includes the cell itself and its membrane system. In fact, we don't need the concept of a program here. The sequence of events, including the feedback between the cell potential and the activity of the proteins, simply *is* cardiac rhythm. It is a property of the interactions between all the components of the system. It doesn't even make sense to talk of cardiac rhythm at the level of proteins and DNA.

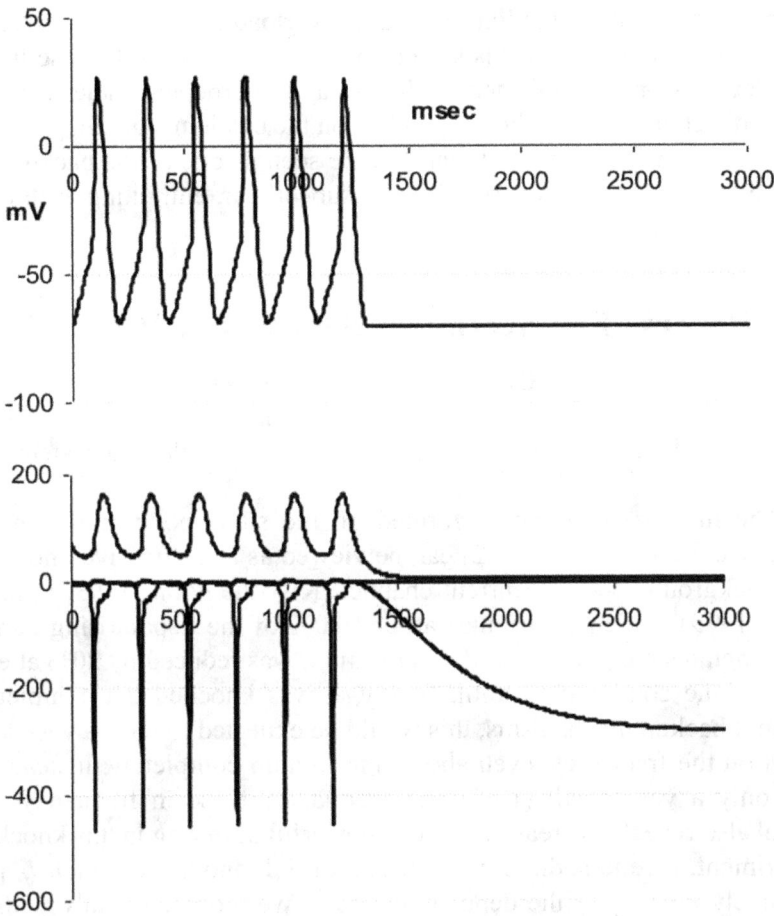

Figure 6.1. Computer model of sinus node pacemaker rhythm (using equations from Noble and Noble, 1984). Three of the protein channels are shown: a potassium channel, a calcium channel and a non-specific (sodium and potassium) channel. After 6 cycles, the feedback from the membrane potential onto the ionic channels was removed by holding the potential constant. All the oscillations disappear and the activity of each channel relaxes to a steady state.

It is important to clarify what is happening here. The demonstration that the rhythm is a property of the complete cell, not of individual proteins or gene–protein networks, is an empirical discovery. Subcellular oscillators can and do also exist, for example, the intracellular calcium oscillator discussed in Chapter 5. In other circumstances, for example, in ischemic conditions (sodium-calcium overload), the model used in Fig. 6.1 would also show this kind of oscillation. That kind of oscillation also depends on 'complicated feedbacks and interdependencies', though the feedback involved is then dependent on the calcium signalling system, the sarcoplasmic reticulum. It can also be seen as one of the backup systems, with each oscillator, cellular and subcellular, entraining each other (Noble *et al.*, 2010).

6.5. Reverse Engineering of the Cardiac Pacemaker

We are now ready to fully appreciate the Noble *et al.* (1992) article. In addition to the existence of two types of oscillators, cellular and subcellular, the cellular oscillator itself is also a good example of a system with backup.

The main experiment performed on the sinus node model in this article (reproduced in Fig. 6.2) can be viewed as a progressive knockout. The background sodium current channel (Kiyosue *et al.*, 1993; Spindler *et al.*, 1999), which contributes around 80% of the depolarising current that generates the pacemaker depolarisation, was reduced by 20% at each stage of the computation until, finally, it was knocked out completely. Without backup mechanisms, this would be expected to have a very large effect on the frequency, even abolishing rhythm completely. Instead, we find only a very small (10%) and gradual decrease in frequency. The model also reveals the reason for such powerful buffering in this knockout experiment. As the sodium channel is reduced, another channel, i_f, progressively takes over the depolarising role. We move smoothly with no interruption of rhythm from a mechanism primarily dependent on one channel to a mechanism primarily dependent on the other. Although the computation doesn't show it, we can also carry out the reverse knockout. Progressive reduction of i_f in a model in which it is the major mechanism also leads to only a modest reduction of frequency as the background sodium channel takes over.

What conclusions can we draw from this model experiment?

First, the insights involved, all the way from the identification of the i_f channel described in Chapter 3 through to the demonstration that its inhibition could be a safe way to reduce heart frequency, led to the realisation by the drug company, Servier, that a specific blocker of the i_f channel could be a very useful drug since slowing cardiac rhythm helps patients

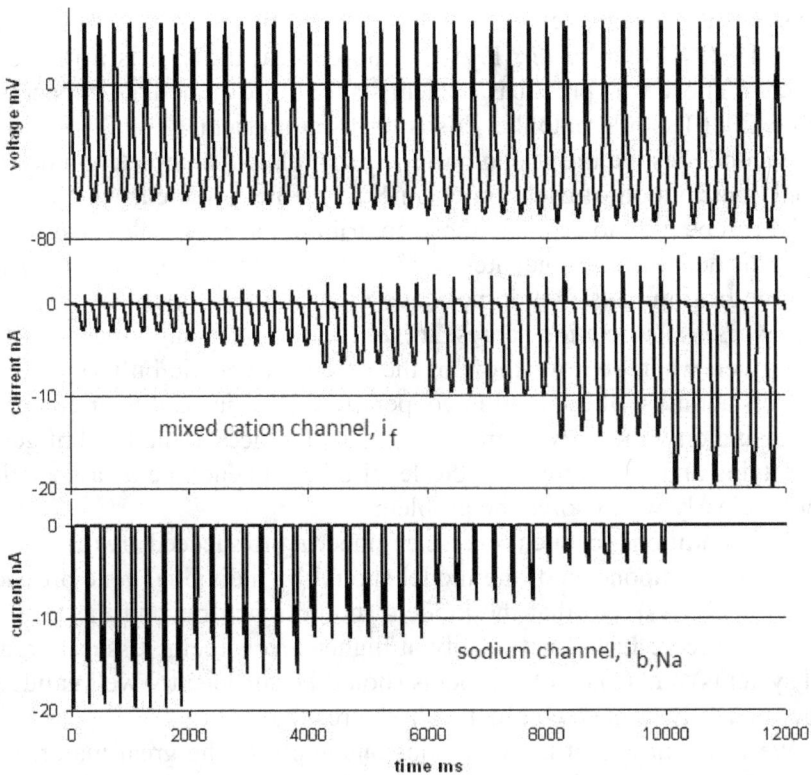

Figure 6.2. Example of the use of computational systems biology to model a genetic buffering mechanism. Top: membrane potential variations in a model of the sinus node pacemaker of the heart. Bottom: the background sodium channel, i_{bNa}, is progressively reduced until it is eventually 'knocked out'. Middle: the mixed (sodium and potassium) cation current channel, i_f, progressively takes over the function, and so ensures that the change in frequency is minimised (recomputed using COR from Noble *et al.*, 1992). Coordinates: membrane potential in mV, current in nA, time (abscissa) in ms.

with ischemia and other forms of heart trouble by reducing the energy demand by the heart. But, of course, the slowing must be limited in range. This is precisely what a block of i_f can achieve. Ivabradine is the drug that was developed and has now achieved FDA and other regulatory approval for use in patients with ischemia. The clinical trials have shown effectiveness, particularly in patients with high heart rates. *The Guardian* featured an article (Pidd, 2010) highlighting the life-saving features of the drug as

I was writing this chapter in 2010. From the initial discovery of the channel in 1979 (Brown *et al.*, 1979) to the clinical trials, it is a period of 30 years. Those who press basic scientists for more immediate impact (see Noble, 2010) should note that this is far from unusual.

Second, despite the strong buffering that almost completely hides the actual contribution of each ion channel to the net ionic current flow, the model can be used to estimate those contributions in a highly quantitative way. This demonstrates the integrative power of reverse engineering using physiological models. We do not have to be limited simply to observing *differences* as association scores. It is difficult to see any other way forward given what we now know of the extent of genetic buffering. Gene products act most of the time in cooperation. Working out their contributions using only the forward mode (using differences at the level of genes or proteins and their effects at the level of the phenotype as association scores) clearly won't solve the problem.

The conditions for such reverse engineering to succeed are (a) that the elementary components of the model should be individual gene products (proteins, RNAs); (b) that the models should reach beyond the level of proteins to reproduce functionality at higher levels (cells, tissues, organs and systems) and (c) that the models should be sufficiently well validated experimentally to have confidence in the results.

We now know that these conclusions apply to the great majority of genome-wide association studies. 95% of genes show very low association scores. That is the fundamental reason why even the summed (polygenic) scores are not reliably predictive of major fatal diseases (Hingorani *et al.*, 2023).

Sadly, integrative physiologists were like lone voices in the wilderness in pointing out this problem many years ago. The balance of biological research and its funding had already shifted so completely that the integrative physiological research that now needs to come to the rescue of the failure of genomics was severely under-funded.

The so-called 'Central Dogma' of molecular biology has a lot to answer for. It should never have been presented as 'dogma', even as a joke. The issues are too serious. The fact that the DNA > RNA > protein process only runs one way cannot possibly prevent the organism from changing its genes when it needs to do so. It doesn't need to do that by reversing that sequence. The immune system was precisely doing genome editing through processes at a very tiny region of the complete genome when it produced new immunoglobulin proteins during the COVID pandemic. McClintock (1984) warned against excluding genome editing by organisms under stress, which she showed triggered chromosome rearrangements. So also has James Shapiro (1983, 2022). There is no 'Central Dogma' therefore that could prevent organisms from editing their genomes (Noble and Noble, 2023), nor does the boundary between the soma and germline prevent control RNAs and transcription factors from crossing it (Phillips and Noble, 2024). There are many cellular and systems-level processes by which organisms can edit their genomes, all of which depend on the fact that DNA is not a *self*-replicator (see Chapter 10 of this book). Its highly accurate replication is itself a function of the living cell, not its DNA.

6.6. Conclusions

6.6.1. *Pluses*

The 1992 article largely resolved arguments about which channels play the largest roles in pacemaker activity by showing that, depending on the conditions, they can replace each other as backup mechanisms. The cardiac pacemaker, just like other biological oscillators such as the circadian rhythm, is a multi-process fail-safe system. The precise contributions of the different mechanisms can vary depending on the conditions under which the heart's pacemaker is operating.

6.6.2. *Minuses*

So far, there are very few models that satisfy all the conditions for reverse engineering to be successful in quantifying the relative contributions of particular gene products. It may seem surprising that this should be so when so much experimental data are available on genes and proteins. The problem lies in the sizes of the networks involved. In the case of cardiac electrophysiology, we have been fortunate in dealing with relatively small networks of proteins and other components that function as a module in relative isolation. By contrast, metabolic pathways and signalling networks often contain hundreds of components. A combinatorial explosion then thwarts even the cleverest analysis. Of course, the electrophysiological networks we have analysed sit 'on top' of those metabolic and signalling networks, so their 'isolation' is only relative. As we expand the models to incorporate those networks, we will face the same kind of difficulty. I see this as one of the greatest challenges facing systems biology. I have not yet read a convincing strategy for meeting this challenge.

6.6.3. *Contribution to systems biology*

For me, at least, this work opened the way to appreciating the fundamental difference between the differential and integral approaches to gene–phenotype relations (Noble, 2011a, 2011b). The full significance of this difference will be explained in Chapter 10.

References

Ball, P. (2024) *How Life Works: A Users Guide to the New Biology*. Basingstoke: Picador. ISBN: 9781529095982.

Brown, H. F., DiFrancesco, D. and Noble, S. J. (1979) 'How does adrenaline accelerate the heart?', *Nature*, 280, pp. 235–236.

Dawkins, R. (1976, 2006) *The Selfish Gene*. Oxford: Oxford University Press.

Dawkins, R. (1982) *The Extended Phenotype*. London: Freeman.

DiFrancesco, D. and Noble, D. (1985) 'A model of cardiac electrical activity incorporating ionic pumps and concentration changes', *Philosophical Transactions of the Royal Society B*, 307, pp. 353–398.

Editorial. (2010) 'The human genome at ten', *Nature*, 464, pp. 649–650.

Hillenmeyer, M. E., Fung, E., Wildenhain, J., Pierce, S. E., Hoon, S., Lee, W., Proctor, M., St Onge, R. P., Tyers, M., Koller, D., Altman, R. B., Davis, R. W., Nislow, C. and Giaever, G. (2008) 'The chemical genomic portrait of yeast: Uncovering a phenotype for all genes', *Science*, 320, pp. 362–365.

Hingorani, A. D., Gratton, J., Finan, C., *et al.* (2023) '*BMJMedicine*', 2, p. e000554. doi:10.1136/bmjmed-2023-000554

Kiyosue, T., Spindler, A. J., Noble, S. J. and Noble, D. (1993) 'Background inward current in ventricular and atrial cells of the guinea-pig', *Proceedings of the Royal Society B*, 252, pp. 65–74.

Kohl, P., Crampin, E., Quinn, T. A. and Noble, D. (2010) 'Systems biology: An approach', *Clinical Pharmacology and Therapeutics*, 88, pp. 25–33.

McClintock, B. (1984) 'The significance of responses of the genome to challenge', *Science*, 226, pp. 792–801.

Noble, D. (2002) 'Is the Genome the Book of Life?' *Physiology News*, 46, pp. 18–20.

Noble, D. (2006) *The Music of Life*. Oxford: Oxford University Press.

Noble, D. (2008) 'Genes and causation', *Philosophical Transactions of the Royal Society A*, 366, pp. 3001–3015.

Noble, D. (2010) 'Funding the pink diamonds: A historical perspective', *Notes and Records of the Royal Society*, 64, pp. 97–102.

Noble, D. (2011a) 'Differential and integral views of genetics in computational systems biology', *Journal of the Royal Society Interface Focus*, 1, pp. 7–15. doi:10.1098/rsfs.2010.0444.

Noble, D. (2011b) 'Neo-Darwinism, the Modern Synthesis, and selfish genes: Are they of use in physiology?' *Journal of Physiology*, 589(Pt 5), pp. 1007–1015. doi:10.1113/jphysiol.2010.201384.

Noble, D. (2024) 'Genes are not the blueprint for life', *Nature*, 626(7998), pp. 254–255. doi:10.1038/d41586-024-00327-x.

Noble, D. and Noble, S. J. (1984) 'A model of S.A. node electrical activity using a modification of the DiFrancesco-Noble (1984) equations', *Proceedings of the Royal Society B*, 222, pp. 295–304.

Noble, D., Denyer, J. C., Brown, H. F. and DiFrancesco, D. (1992) 'Reciprocal role of the inward currents $i_{b,Na}$ and i_f in controlling and stabilizing pacemaker frequency of rabbit sino-atrial node cells', *Proceedings of the Royal Society B: Biological Sciences*, 250, pp. 199–207.

Noble, D., Noble, P. J. and Fink, M. (2010) 'Competing oscillators in cardiac pacemaking: Historical background', *Circulation Research*, 106, pp. 1791–1797.

Noble, R. and Noble, D. (2023) *Understanding Living Systems*. Cambridge: Cambridge University Press ISBN: 9781009277365.

Phillips, D. and Noble, D. (2024) 'Bubbling beyond the barrier: Exosomal RNA as a vehicle for soma-germline communication', *Journal of Physiology*, 602(11), pp. 2547–2563. doi:10.1113/JP284420.

Pidd, H. (2010) 'Heart failure pill "could save thousands of lives"', *The Guardian*, 30 August 2010.

Shapiro, J. A. (1983) 'Variation as a genetic engineering process', In *Evolution from Molecules to Men* (ed. D. S. Bendall), pp. 253–270. Cambridge: Cambridge University Press.

Shapiro, J. A. (2022) *Evolution: A View from the 21st Century. Fortified. (2nd Edition)*. Chicago IL, USA: Cognition Press.

Spindler, A. J., Noble, S. J., Noble, D. and LeGuennec, J. Y. (1999) 'The effects of sodium substitution on currents determining the resting potential in guinea-pig ventricular cells', *Experimental Physiology*, 83, pp. 121–136.

Chapter 7

The Physiome Project

7.1. The First Massively Parallel Computers

In 1989, I was invited to the University of Minnesota to give a lecture at the Bacaner Research Awards ceremony. Marvin Bacaner, a physiologist at the University since the 1960s, was responsible for the development of the anti-arrhythmic drug, bretylium, which works by releasing noradrenaline and then inhibiting further release from nerve endings. He had given the University some of the substantial royalties from bretylium to establish the prizes and lecture, and later (in 2000) gave a further $0.5 million to help establish a chair at the University.

My lecture described the development of cellular models of the heart. This was shortly after the development of the DiFrancesco–Noble, Hilgemann–Noble and Earm–Noble models. Interest in using these and in extending them was already high around the world. After describing these developments at the cellular level, I ended the lecture by lamenting that we would have to wait until the 21st century to see the use of such models in reproducing activity at the levels of multicellular tissue and the whole organ. I was guessing that it would take that long for computational power to reach the required level. During the discussion, though, I was told by Rai Winslow, then at Minnesota but later at Johns Hopkins University, that I was wrong. The computing power was already there in the form of the US Army High-Performance Computer installation coming online at the US Army High-Performance Computing Research Center at the University of Minnesota. This machine was a giant in those days. Called the Connection Machine CM-2 and produced by Thinking Machines Inc., it achieved its power by having 64,000 processors ganged together in a grid (Hillis, 1989).[1] Connection Machine was the right name for it.

Rai had already seen that this computer was ideal for some of the tissue-level simulations he had in mind. He had already used it to develop biophysically detailed models of the horizontal cell network in the retina. The structure of the machine, with each processor communicating directly with the neighbouring processors, mirrors the structure of heart tissue, with each cell communicating by electrical connections (the nexus protein junctions) to its neighbours. You could therefore map a block of tissue onto the processors themselves, in the simplest case by assigning a cell model to each processor, so generating a block of tissue corresponding to 64,000 cells.

So, I agreed to provide input on the cell models for the study he proposed. For me, this was the beginning of the Physiome Project (Bassingthwaighte *et al.*, 2009; Noble, 2002), also beginning on the other side of the world in the laboratory of Peter Hunter in Auckland. I will return to Auckland later in this chapter. Three very interesting results came out of the Connection Machine work with Rai and his colleagues (Winslow *et al.*, 1991a, 1991b, 1991c).

The first was an answer to a question I had been interested in for a long time, in fact, since a symposium in Holland (Bouman and Jongsma, 1982) at which Robert deHaan (DeHaan, 1982) presented a remarkable result. Two chick ventricle heart cells synchronised their rhythms with, apparently, almost no connection between them! Actually, there were long, thin projections between the cells. But, even if these touched, there wouldn't be room for more than a nexus junction or two. DeHaan's electron microscopic images showed such a nexus at the very end of the process (DeHaan, 1982, Fig. 4). Could that single nexus conceivably synchronise two cells? During his talk, I did a quick back-of-the-envelope calculation, starting with the net current flowing in a single cell during the pacemaker depolarisation. This is very small, just a fraction of a picoamp. Then, I asked the question of how much current could flow through a single nexus with a voltage difference of just a few mV, knowing the nexus conductance. The calculations matched (Noble, 1982)! Although I could not prove it, the result showed that it was plausible that even a single nexus could have allowed the two cells to synchronise.

---

[1] Remarkably, the Connection Machine was Danny Hillis's PhD thesis project with Tomasso Poggio at MIT. For a while, Thinking Machines Inc. was the leading supercomputer vendor in the world, exceeding Cray in sales.

The proof had to wait for the Connection Machine simulations. We took cell models of the sinus node with different intrinsic frequencies and connected them together with different levels of conductance between the cells. Even four or five gap junctions with a conductance of 50 pS were sufficient to allow the complete network to synchronise (Winslow *et al.*, 1992). At that time, we also knew that the connexin 43 (the protein that forms the nexus junctions) density in the sinus node was very low compared to the atrium and ventricle. The electrical connections between sinus node cells are indeed weak compared to those between atrial or ventricular cells. Problem solved (Cai *et al.*, 1994).

The second problem we tackled involved introducing known spatial variations in cell properties within the sinus node. When isolated, cells at the centre of the node show a relatively small voltage variation and a lower frequency, whereas cells at the periphery display larger voltage changes and a somewhat higher frequency. The question was, 'What determines where the impulse starts in the node?' In the intact heart, the impulse usually starts at or near the centre of the node, although the precise location changes with autonomic nerve and hormonal activity (Meek and Eyster, 1914; Bouman *et al.*, 1968; Shibata *et al.*, 2001).

The computation showed precisely the wrong result (Winslow *et al.*, 1993). The initiation of the impulse started at the periphery and was conducted towards the centre. I shall never forget the first time I presented the lovely film that Rai and his team produced of this result at a meeting in the UK. I simply showed it as an example of what could be achieved in multilevel simulation using the Connection Machine, but admitted that there was a problem with the result and that we would be investigating ways in which we could understand that and what would be required for the initiation to begin towards the centre.

Fortunately, Mark Boyett (then working in Leeds) was in the audience. His comment was dramatic. 'You don't need to look any further. The result is correct!' I am ashamed to say that I did not know of the work he had done on this question, but his experiment was precisely what was needed. He had shown that if you carefully cut around the sinus node to isolate it from the atrial tissue, the impulse does indeed start at the periphery. The atrial cells effectively suppress the pacemaker depolarisation in the periphery by supplying hyperpolarising current. Boyett and his colleagues subsequently analysed this phenomenon in more detail than we did (Toyama *et al.*, 1995).

The third major contribution from the work on the Connection Machine was designed to answer a question that is relevant to the mechanisms of arrhythmia. One of the advantages of the Hilgemann–Noble and Earm–Noble models is that they can reproduce the spontaneous depolarisations in sodium-overloaded conditions that are attributable to cycling of the intracellular calcium oscillator. In conditions where intracellular sodium is high, intracellular calcium also increases since the sodium gradient driving the calcium efflux through the sodium–calcium exchanger is weakened. This is the mechanism that was first studied by Dick Tsien working with Jon Lederer (see Chapter 5). An important question is how extensive the sodium-overloaded tissue must be in order to generate a propagated extra action potential. This was an ideal question for the Connection Machine since, once again, a sheet or block of cells could be modelled in an efficient way. The answer was that the region of overloaded cells does not have to be very large. In a network of about a quarter of a million cells, only about 1000 cells with sodium overload can induce a propagating ectopic beat. The explanation for this result is that the sodium–calcium exchanger is very effective as a current generator. Since the current it carries rapidly increases as the membrane is hyperpolarised (see Fig. 5.3 in Chapter 5), the mechanism resists hyperpolarisation. This result could be relevant to questions concerning arrhythmic mechanisms in ischemia. The ischemic region need not be very extensive in order to generate an arrhythmia.

7.2. Auckland and the Virtual Ventricle

Collaboration with Peter Hunter goes back to the 1970s, when he was in Oxford and we worked on some of the analytical mathematics of excitation and conduction in nerve and muscle described in Chapter 3.

My first visit to his laboratory in Auckland was in 1990 when I was invited to be the Butland Visiting Professor. My journey there was almost as memorable as the flight on the Concorde since it is the only time I have flown right across the world in first class. I arrived at the Air New Zealand desk at Heathrow to be informed that the flight was full and that, 'Sorry, sir', I couldn't be fitted in! Initially, that was all I heard. My mind was already on the question of how to communicate to my hosts that I would, presumably, be coming at least a day later. Then, I noticed that I was being given the boarding pass saying first class and the invitation to the lounge. I had been booted up from economy to first class. That was my introduction to a top New Zealand white wine as I was pampered and cosseted all the way to Los Angeles and then on to Auckland. Actually, it was far better than the Concorde experience. To have a completely flat bed (only in first class in those days) on such a long journey was a miracle.

I arrived early in the morning to be ushered immediately into the laboratory where Bruce Smaill, Peter Hunter and their collaborators were developing an accurate computer model of the anatomy of the ventricle. I am not sure I paid as much attention as I should have to what they were doing. The problem was not the jet lag (as explained in Chapter 1, I don't find that a big problem). It was rather that the Marlborough Sauvignon Blanc, from the South Island, served to the first-class passengers, had made a deep impression on me, probably more than it should have. If you appreciate top-quality Sancerre, the nearest French equivalent, you will know what I mean. One respected wine critic has described Marlborough Sauvignon Blanc as 'having sex for the first time' (Taber, 2005, page 244). I am not sure I agree with that; Naomi Wolf (1998, page 133) asks, 'Is that it?', echoing many people's first experience disappointment – of sex, not wine! Anyway, my first experience of a top New Zealand white was simply magical.

What a way to begin what was to become the Physiome Project!

So, what was the Auckland team doing? They were using stereotactic instruments to computerise a whole heart by progressively shaving off very thin layers and recording the fibre orientations over the entire exposed surface, until a complete three-dimensional image was created in the form of a database. The idea was that this database could eventually be combined with equations for the mechanical, electrical and biochemical properties of the cells in each region of the heart to produce the world's first virtual organ. The critical characteristic of the database was that it included information on the fibre orientation since this determines both the direction in which force is generated when the cells contract and the preferential route for conduction of the electrical impulse.

The Auckland team was also developing the mathematical framework for characterising this tissue structure in a way that could be used in the solution of the governing physical laws that determine organ behaviour. Most engineering structures (cars, bridges, aircraft, etc.) are designed using a numerical technique called the 'finite element method' that considers a continuous material as a collection of small regions ('finite elements') linked together through points ('nodes') on their boundaries. It is a 'divide-and-conquer' strategy that is well suited to dealing with complex geometries and inhomogeneous (spatially varying) material properties on digital computers. The geometry and material structure of biological organs such as the heart are of course highly complex, so it is no surprise that these techniques developed for the world of engineering should be equally relevant to the world of physiology. There is, however, a very close relationship between organ geometry and tissue structure that is not often seen in man-made engineering structures. The particular finite element methods developed by the Auckland group in the 1980s were designed to capture this close relationship and have now become the basis for much subsequent physiome modelling of the body's organ systems.

This work beautifully complemented the tissue-level work that was being done in collaboration with Rai Winslow. But the computational demands created by simulating at the level of the whole organ were even larger than for simulating multicellular blocks of tissue. The simulations described in Chapter 4 that were done by Rai Winslow and his team to reconstruct the T wave of the electrocardiogram illustrate the problem. It took around 6 weeks of work on a supercomputer to complete that project. It is only during the last decade or so that we have been able to harness enough computing resources to deal effectively with the challenges of whole-organ simulation.

It should be noted also that even the most complete of the cell models of the heart do not represent more than a small fraction of the total number of components, i.e. genes, proteins, metabolites and membrane structures, involved. Certainly, we should be using whatever computing power comes our way as the speed and capacity rapidly increase. But I think it is also likely that the practical and theoretical limits to this approach will force us to consider alternatives to modelling the body. I will return to this question in Chapter 9.

7.3. The Physiome Project of the International Union of Physiological Sciences (IUPS)

In 1993, I was responsible as Chairman of the Organising Committee for arranging the World Congress of Physiological Sciences held in Glasgow. Nearly 5,000 people attended the meeting, possibly the largest IUPS Congress ever to be held. We used the opportunity to publish a book, *The Logic of Life* (Boyd and Noble, 1993), which was distributed to all members of the Congress. I see that book as an early precursor of what was to lead eventually to the Physiome Project and to the physiological aspects of systems biology. It was also the Congress at which Sydney Brenner gave a plenary lecture in which he challenged physiologists to respond to the need to interpret the genome.

My colleagues and I met after the Congress to discuss how to react to that challenge. By then, I also knew that I was to become the next Secretary-General of IUPS. I served in that role from 1994 to 2001, with two World Congresses occurring during that period. The first was at St Petersburg in 1997. By that time, Jim Bassingthwaighte in Seattle had already proposed the idea of the Physiome Project. The ending – ome – means 'totality', of course. Genome refers to the complete set of DNA sequences of an organism. Physiome therefore means the totality of physiological function in an organism. That is precisely what physiology is, or should be. Why, therefore, invent a different word?

The main reason, I think, was to widen the scope of the project well beyond the discipline of physiology itself to include the bioengineers, mathematicians and computer scientists, amongst others from the physical sciences, who would need to bring their skills to bear on what is essentially a multi-disciplinary project, to reconstruct the organism mathematically. We did not know then that, within a few years, similar ideas on the role of mathematics and computer modelling in biology would develop within the context of systems biology. I will discuss the relationships between systems biology, the Physiome Project and the Virtual Physiological Human project later in this chapter.

The formal launch meeting of the Physiome Project was held just after the St Petersburg Congress. About 40 of us stayed at a beautiful but rundown former palace on the coast to the west of the city, near Petrodvorets, where the famous Peterhof Palace is located. The outcome was the formation of the Physiome Committee of IUPS, with Peter Hunter from Auckland as chair and Sasha Popel from Johns Hopkins University as co-chair. This committee has run very successful symposia and satellite meetings at all the subsequent international congresses, beginning with the Congress in Christchurch, New Zealand, in 2001. The project has moved from small beginnings to see substantial funding, particularly in the USA, UK and Japan, while the EU has launched the related Virtual Physiological Human (VPH) project.

In the early days, however, funding was not easy. This was one of the reasons why Rai Winslow, Peter Hunter and I were involved for several years in the development of a venture capital company, *Physiome Sciences Inc*, which attracted substantial funding both for the research and for its exploitation. In its last round of funding, it raised around $50 million in capital investment and had a substantial facility located in Princeton, New Jersey. The work done for Roche (Chapter 4) was an early success for exploitation and encouraged us to think that a viable business model was possible. The involvement of many large pharmaceutical companies in the publicly funded EU successor to the project, preDiCT (Chapter 4), shows that this may well have been true. Two factors, however, ensured that this venture capital approach did not survive. One was that we probably underestimated the time it would take to develop the work to a fully exploitable stage. The second was the catastrophic effect of the events of 9/11 on the venture capital biotech market. Quite simply, the sentiment of the market completely switched after the fall of the Twin Towers.

There are, however, some lasting gains that are attributable to *Physiome Sciences*. It funded important work on the Physiome Project at a time when public funding was not available on the scale required. The Johns Hopkins, Oxford and Auckland teams benefited substantially from that funding. And it raised awareness of the potential for applications of biological modelling work in the pharmaceutical world. Some of the subsequent collaborations, particularly with Novartis, Roche and Pfizer, owe their origins to the work of *Physiome Sciences*.

Possibly the most important open public legacy of the *Physiome Sciences* era was the CellML standard. Very few mathematical models of biological processes are published in peer-reviewed scientific journals without either errors (typographical or otherwise) or missing parameter values. The more complex the model, the more difficult it is to reproduce the published results from the equations laid out in the publication. One strategy for addressing this problem is to include computer code on a website, but even that is often unsatisfactory because the equations implemented in the code may well not exactly match the ones written down in the paper itself. The only solution is to code the mathematics in an unambiguous 'marked up' form, rather like the HTML standard used for web pages. CellML is an XML (eXtensible Markup Language) framework for specifying the equations, together with their units and information about the model structure. It provides the syntax or 'grammar' for the model, while the biophysical and biological meaning – the semantics – of variables and parameters is specified with ontologies via the CellML metadata standard. The equations and computer code for a model can be generated automatically from the CellML syntactical encoding, and the ability to combine models as components of more complex models is facilitated by the semantic metadata. The CellML standard was developed by a team from Auckland University and *Physiome Sciences*, primarily under the leadership of Poul Nielsen from Auckland. A related but independent effort at Caltech on SBML (Systems Biology Markup Language) began shortly after. These two efforts and a third one from Auckland around the 'FieldML' standard for encoding spatial information are now at the heart of the Physiome Project and are providing a robust foundation for biological modelling.

7.4. The Coming of Systems Biology

When I first wrote this chapter in 2010, there were nearly 3,000 papers published with the phrase 'systems biology' in their titles or abstracts. Of these, only two were published before 2,000 and 90% were published in the five years after 2005. What has happened? And how does the Physiome Project relate to this sudden development?

One explanation for the development of 'systems biology' as an idea is that the Human Genome Project, completed in draft form at the turn of the millennium, was the natural culmination of the reductionist drive to burrow right down to the smallest elements in an organism, the DNA sequences that are responsible for transmitting genetic information from one generation to its successors. Having reached this level, though, where next could the reductionist agenda take us? It might be plausible to imagine, as some do, that the whole of life might be seen to be encoded in DNA (needless to say, I am not amongst those who think this way), but it is clearly implausible to suppose that it can be encoded, in anything other than an extremely general sense, below that level. I suppose that strict determinists would need to postulate that life is indeed inherent in the properties of matter even at the level of, say, fundamental particles. But that is a very general sense of 'inherent' that is, at present at least, of no scientific value. We have no way either of coherently formulating such a hypothesis or of submitting it to any kind of experimental test. In the reductionist analysis, DNA really is the bottom level with which we have to work today.

Having reached that level, do we now understand life?

In *Science Oxford*, a public forum for science, where I gave a lecture in 2009, I spent some time looking at a huge book on display in the exhibition area. It consisted of the printing of the complete sequence of the DNA in a single chromosome. Occasionally in the mammoth book, as you turn the pages, you can see a highlighted section that is known to correspond to a gene. For those sections, the triplets of nucleic acids corresponding to each amino acid would tell you for which protein the sequence forms a template. Huge sections though are just, apparently, gibberish code for which we have only the glimmering of a 'meaning'. Is this book really 'the book of life'? At the least, we have to admit that from being able to scan it in the casual way I was doing in *Science Oxford*, no sudden enlightenment arises. On the contrary, one reads this much as one would read the gibberish binary data that might correspond, on my computer, to the text I am currently writing. At that level, one would need an interpretative program – called Word on my PC – which is of course the level at which the vast majority of us operate. It is part of the purpose of disciplines like bioinformatics to be able to do that with the DNA sequences of the genome.

Undoubtedly, bioinformatics of this kind has made substantial progress in comparing genomes from different species to arrive at some fundamental conclusions about evolution and the relevant trees of life. We need that kind of study of biological systems and how they have developed. But there is a very simple reason why, from the sequences alone, we will not understand life. This is that those sequences do absolutely nothing until they receive signals – transcription factors and epigenetic marking of various kinds – from the rest of the organism, starting in development with the fertilised egg cell and its environment (the mother's womb in the case of mammals). If we wish to attribute meaning to something, it must surely be in those signals. They 'command' the genome to the only act of which it is capable, which is transcription. Well, even that is not strictly true – that act is performed by a dedicated set of proteins working on the DNA sequences.

That is precisely the realisation that leads to the main idea of systems biology. To understand those signals acting on the genome, you have to understand the system of which they are a part. Systems biology can therefore be seen to be 'bouncing back' to higher levels in the organism itself after having reached the bottom in sequencing its genome.

But that is where the trouble begins. Where in the multilevel system that is an organism do you bounce back to? Some, like Sydney Brenner, would insist that you have to bounce back to the level of the cell. Others might claim that metabolic and developmental pathways will do the trick. Certainly, over 90% of what is called systems biology today is largely at the level of biochemical pathways and the analysis of huge amounts of genomic and proteomic data. Very little of what is being done under its name lies above the cell level.

One of the principles of systems biology that I formulated in an article in 2008 (Noble, 2008) is what I call the principle of biological relativity. It states that, a priori, there is no privileged level of causality in biological systems. In an organism in which there are multiple feedbacks and feed-forwards between all the levels from the genome to the complete pheno-type, and even beyond to its environment, I would say that this insight is a necessary one. The emphasis in what I have just written must be on the 'a priori'. Before investigating a system, we cannot say at which level a particular phenotype is integrated and therefore understood. That doesn't mean to say that in investigating a particular phenotype we can't discover that, as a matter of fact, it is expressed or integrated at a particular level. Pacemaker activity of the heart, for example, can now be seen to be a function integrated at the level of a cell, though it is also greatly modu-lated by even higher levels.

It is in this question of level of causality that we can understand the relation between the Physiome Project and systems biology. The Physiome Project and its relative, the Virtual Physiological Human (see in the fol-lowing), complete the intellectual revolution in biology that systems biol-ogy represents. Without them, there is no guarantee that systems biology can succeed. The one is an essential tool for the other.

7.5. The Virtual Physiological Human (VPH)

The systems approach to biology is being driven by many different streams of activity. One of those streams is the world of computing. As the Physiome Project has developed, it has rapidly become one of the grand challenges for ever more powerful supercomputers, and more recently for cloud computing harnessing the power of many computers. If you want a test-bed for your nation's latest mammoth machine, biology provides that in abundance. So much so that it has become one of the drivers of super-computing itself. This is the origin of the European Union's flagship project in this area, the Virtual Physiological Human.

Launched under EU Framework 7, this project, or rather set of proj-ects, is funded by the Information and Communications Technology (ICT) section, and is therefore a good example of this area of technology acting as a major driver of work on systems biology focused at the higher levels of function.

7.6. Conclusions

7.6.1. *Pluses and contributions to systems biology*

The development of the Physiome Project is essential to the success of the systems biology approach. This is appreciated by some of the major players in systems biology. When I first drafted this chapter, I took part in a debate with Hans Westerhoff, who has successfully championed the systems approach worldwide in many ways. The strange situation is that, although his own focus has been on the systems biology of yeast, he was arguing *for* the Virtual Human, while I was acting the part of the sceptic! That is how it should be. An approach that promises everything 'in a decade or two' will fail. To avoid the publicity over-reach of the Human Genome Project, we need the debate, and we all need to be sceptical. Science thrives on scepticism.

7.6.2. *Minuses*

I am not sure, however, that there are many like Hans Westerhoff in the systems biology camp. I would like to be proved wrong. Part of the problem is that, with the decimation of classical physiology during the decades of dominance by the molecular biological approach, we don't seem to have many world leaders left to argue the case. Or perhaps people are, naturally, more concerned with their next grant request? A major reason why I accepted the nomination to become the President of the International Union of Physiological Sciences (IUPS) in 2009 and held that position until 2017 is that I see outreach towards our fellow scientists and the general public as a major role that an international union should be playing. This is also a suitable point at which to acknowledge the role of bioengineering in 'keeping the faith'. It is no accident that many of the major proponents of the Physiome Project, Jim Bassingthwaighte, Peter Hunter, Sasha Popel and Rai Winslow, are all bioengineers and that many graduate students and postdoctoral fellows who are joining the work come from the physical sciences.

References

Bassingthwaighte, J. B., Hunter, P. J. and Noble, D. (2009) 'The cardiac physiome: Perspectives for the future', *Experimental Physiology*, 94, pp. 597–605.

Bouman, L. N., Gerlings, E. D., Biersteker, P. A. and Bonke, F. I. M. (1968) 'Pacemaker shift in the sino-atrial node during vagal stimulation', *Pflügers Archiv, European Journal of Physiology*, 302, pp. 255–267.

Bouman, L. N. and Jongsma, H. J. (1982) *Cardiac Rate and Rhythm*. The Hague: Martinus Nijhoff.

Boyd, C. A. R. and Noble, D. (ed.) (1993) *The Logic of Life*. Oxford: Oxford University Press.

Cai, D., Winslow, R. and Noble, D. (1994) 'Effects of gap junction conductance on dynamics of sinoatrial node cells: Two-cell and large-scale network models', *IEEE Transactions on Biomedical Engineering*, 41, pp. 217–231.

DeHaan, R. L. (1982) 'In vitro models of entrainment of cardiac cells', In *Cardiac Rate and Rhythm* (ed. Bouman, L. N. and Jongsma, H. J.), pp. 323–359. The Hague: Martinus Nijhoff.

Hillis, D. (1989) *The Connection Machine*. Cambridge, Mass: MIT Press.

Meek, W. J. and Eyster, J. A. E. (1914) 'The effect of vagal stimulation and of cooling on the location of the pacemaker within the sino-auricular node', *American Journal of Physiology*, 34, pp. 368–383.

Noble, D. (1982) 'Discussion', In *Cardiac Rate and Rhythm* (ed. Bouman, L. N. and Jongsma, H. J.), pp. 359–361. The Hague: Martinus Nijhoff.

Noble, D. (2002) 'Modelling the heart: From genes to cells to the whole organ', *Science*, 295, pp. 1678–1682.

Noble, D. (2008) 'Claude Bernard, the first systems biologist, and the future of physiology', *Experimental Physiology*, 93, pp. 16–26.

Shibata, N., Inada, S., Mitsui, K., Honjo, H., Yamamoto, M., Niwa, R., Boyett, M. R. and Kodama, I. (2001) 'Pacemaker shift in the rabbit sinoatrial node in response to vagal nerve stimulation', *Experimental Physiology*, 86, pp. 177–184.

Taber, G. M. (2005) *Judgment of Paris: California vs France and the Historic 1976 Paris Tasting that Revolutionized Wine*. New York: Scribner.

Toyama, J., Boyett, M. R., Watanabe, E., Honjo, H., Anno, T. and Kodama, I. (1995) 'Computer simulation of the electrotonic modulation of pacemaker activity in the sinoatrial node by atrial muscle', *Journal of Electrocardiology*, 28 Supp 1, pp. 212–215.

Winslow, R., Cai, D. and Noble, D. (1992) 'Effects of gap junction conductance on oscillation properties of coupled sino-atrial node cells', *IEEE, Computers in Cardiology 1992*, pp. 579–582.

Winslow, R., Kimball, A., Noble, D. and Denyer, J. C. (1991a) 'Computational models of the mammalian cardiac sinus node implemented on a Connection Machine CM-2', *Medical & Biological Engineering & Computing*, 29, p. 832.

Winslow, R., Kimball, A., Noble, D., Denyer, J. C. and Varghese, A. (1991b) 'Modelling large SA node-atrial cell networks on a massively parallel computer', *Journal of Physiology*, 446, p. 242.

Winslow, R., Kimball, A., Noble, D., Denyer, J. C. and Varghese, A. (1991c) 'Simulation of very large sinus node and atrial cell networks on the Connection Machine CM-2 massively parallel computer', *Journal of Physiology*, 438, p. 180.

Winslow, R., Kimball, A., Varghese, A. and Noble, D. (1993) 'Simulating cardiac sinus and atrial network dynamics on the connection machine', *Physica D: Non-Linear Phenomena*, 64(1–3), pp. 281–298.

Wolf, N. (1998) *Promiscuities: A Secret History of Female Desire*. London: Chatto & Windus.

Chapter 8

50 Years On

8.1. Still at It

November 5th 2010, which was when I first wrote this chapter, marked 50 years since the 1960 articles in *Nature*. I was still modelling cardiac cells, as the articles relevant to this chapter show. The Ten Tusscher model in 2004 was one of the first to incorporate experimental data from work on human cells. The 2007 article was based on a Hodgkin–Huxley–Katz review lecture given for The Physiological Society in 2004 in which I outlined how we progressed from the Hodgkin–Huxley equations of 1952 to the development of a virtual heart Noble, 2007).

There are now well over 100 cardiac cell models on the CellML website, all downloadable and ready for anyone, anywhere in the world to use. They can do so on laptops vastly more powerful than the Mercury computer I used in 1960. The range of species and cell types grows each year. This range of species and cell types has proved valuable in collaborations with pharmaceutical companies who need ways in which we can more reliably extrapolate from common laboratory species like the mouse, rat and rabbit to the human. They also need to resolve the QT problem (the arrhythmic side effects of many drugs) referred to in Chapter 4. Projects being funded by the European Union are tackling that problem (Noble, 2008a; Fink and Noble, 2009, 2010; Garny *et al.*, 2009; Niederer *et al.*, 2009; Stewart *et al.*, 2009; Rodriguez *et al.*, 2010).

Higher-level modelling of tissue, and of the whole organ, has also developed to the stage at which there are teams active around the world. The Cardiac Physiome Project is in good form (Hunter *et al.*, 2006;

Bassingthwaighte *et al.*, 2009). It is also being used as a paradigm for comparable work on other organs and systems of the body, including the kidney, liver and lung.

8.2. Who was He 65 Years Ago?

So, if I could ask the 23-year-old walking down the worn steps to the basement room where the Mercury computer lived in 1960 the question, 'Where did he think this was leading?', what would he have said? Not much probably. How many students see much beyond their thesis when they are writing it? He would have been astonished by the publication 15 years later of *Electric Current Flow in Excitable Cells* (Jack *et al.*, 1975) with 'you don't know enough mathematics!' ringing in his ears and the humbling discussion with Andrew Huxley on Bessel functions of imaginary arguments still to come. He would have been even more astonished by the election to The Royal Society 4 years after that. He certainly would not have anticipated the congratulatory letter from JZ Young, his former Anatomy Professor, mixing the congratulations with the strange comment that he had 'waited rather a long time'. JZ, as he was always called, was elected very young indeed to The Royal Society, at the age of 38.

But what would astonish him most would be the philosophical journey. He was about as reductionist as you could get in 1960: so much so that, intellectually, I hardly recognise him as the same person. His first interactions with philosophers were motivated by the wish to explore that position. He even wrote articles on determinism for one of the UCL journals flourishing at that time. One of his professors, DR Wilkie, a muscle biophysicist, liked it and asked him when the sequel was going to appear. It never did! The reasons are now obvious. He was barking up the wrong tree, but simply didn't know it. A remark by Stuart Hampshire, then Professor of Philosophy at UCL, began the process of finding the right tree. The 23-year-old tried to give an account of action at one of Hampshire's graduate philosophy classes. The argument was based on distinguishing between causes within the organism and causes outside the organism. The response was direct and curt: 'You need to read Spinoza'. The relevant statement is definition VII in the *Ethics*: 'That thing is called free, which exists solely by the necessity of its own nature, and of which the action is determined by itself alone'. Needless to say, Hampshire had just published a book on Spinoza (Hampshire, 1956).

Spinoza was, of course, also the arch anti-dualist. His *Ethics* laid out the grounds for opposition to Descartes's view of the relation between mind and body. I read Hampshire and other material (Elwes, 1951) on Spinoza as a graduate student, but I must have squirreled it all away. For it didn't really play a major overt role in my gradual shift to a completely different philosophical position, despite the fact that Chapters 9 and 10 of *The Music of Life* are strongly non-dualist. It is impossible to exclude a subconscious influence, of course. But what I can be sure of is that the 23-year-old would not have understood the arguments well enough to make effective use of them.[1]

In fact, when I contemplate that person and try to think myself back into his skin, I am surprised at his naivety. Some of my student friends in philosophy at that time must also have wondered what hope there was for this strange mixture of cardiac experimentalist, mathematical modeller and naïve philosopher. He wasn't exactly brilliant at any one of them. He happened to like combining them and that was unique enough to plough completely new ground. Naturally, some of those friends were women. He was greatly influenced by some of them. Notice that my philosopher in Chapter 9 of *The Music of Life* is a woman, so also is the brilliant space-travelling linguist in Chapter 10. In the heavily smoke-filled seminar classes (who nowadays remember that time? I was extremely rare as a non-smoker) given in the philosophy department, many of the students were women. That was the milieu in which the 23-year-old began to grow up. It is hard to imagine those days now. The air of sophistication in those classes seemed to be associated with knowing how to combine making a profound argument with pointed gestures using the ritual of matches and cigarettes. Meanwhile, another physiologist elsewhere in London, Richard Doll at the Central Middlesex Hospital, was busily

---

[1] That has become clear to me in a dramatic way recently. This quotation from one of Spinoza's letters could have been, but wasn't, the basis of my use of the Silman stories in *The Music of Life*: 'Let us imagine, with your permission, a little worm, living in the blood, able to distinguish by sight the particles of blood, lymph etc, and to reflect on the manner in which each particle, on meeting with another particle, either is repulsed, or communicates a portion of its own motion. This little worm would live in the blood, in the same way as we live in a part of the universe, and would consider each particle of blood, not as a part, but as a whole. He would be unable to determine, how all the parts are modified by the general nature of blood, and are compelled by it to adapt themselves, so as to stand in a fixed relation to one another' (letter XV to Oldenburg, 1663).

Figure 8.1.   Denis (left) and Ray Noble singing *Se Canta* – a traditional Occitan song – at a performance of the Oxford Trobadors in the Holywell Music Room, August 2010. This is the concert referred to in the postscript.

refining the bombshell he had published in 1950, showing the link with lung cancer. The 23-year-old absorbed most of the influences of the seminars – and the brilliant women students – except for the cigarettes. His father had died 4 years earlier after many years of chain smoking. He died far too young for a family of four brothers, leaving a single mother to bring them up. As the eldest, the 23-year-old became a substitute father. The youngest of those brothers, Ray, eventually became an academic at University College London, founding the Centre for Reproductive Ethics. Many of the stages in the transition were argued out between those two brothers, who still today have an extraordinary intellectual empathy. The 65-year journey towards enlightenment has involved a great companion (Fig. 8.1).

8.3. The Threads of Transition

But to return to the central puzzle of this book, what led a young reductionist determinist to progressively abandon the naive position he held? The immediate background to writing *The Music of Life* will be dealt with in the next chapter. Here, I am more concerned with the earlier stages.

There were two strands of the process.

8.3.1. *Scientific thread*

One was scientific. While studying ion channels in excitable membranes was near the bottom of the reductionist agenda in the 1960s, the grand sweep of molecular biology soon removed that kind of work from its apparently privileged pedestal. There was a critical period around 1980 when I had to decide whether to follow many of my colleagues down to the molecular level. Something made me hesitate. Yes, it was impressive to see the molecular structure of one type of ionic channel after another being revealed. At last, the molecular basis of, for example, the gating process could be unravelled, and it largely confirmed Hodgkin and Huxley's idea of charged regions of the channel that move in the electric field. In the case of the inactivation gate on a sodium channel, this is literally so: part of the intracellular amino acid chain swings just like a gate to open or close the channel. We can also visualise where the individual ions sit in a channel during the process of transport though the channel. The invention of the patch clamp method, for which Neher and Sakmann received the Nobel Prize in 1991, has enabled us to study the opening and closing of individual channel proteins. All of this has been important in establishing the molecular basis of channel protein behaviour.

The problem I could see was not whether this shift to the molecular level was important or worthwhile. It clearly was. The problem was more sociological. There is a strong herd instinct amongst scientists. There is a gold rush each time a new vista is opened up. For technical reasons, that rush tends to be towards lower and lower levels as instruments become finer and capable of resolving smaller detail. The problem this creates for biological science is that its multilevel nature absolutely requires work at *all* levels. A reductionist would argue that, nevertheless, one starts at the bottom since that has privileged causality. It doesn't (see Chapters 9 and 10). Moreover, we often need the spectacles provided by understanding at higher levels in order to interpret the data at the lower levels.

I am instinctively someone who swims against the tide. I think it was at a meeting on ion channels in the heart at which nearly every talk concerned the same question – analysis of single-channel current jumps as the channels open and close – that I came to the conclusion that, while this was important molecular biophysical work, it was not what I wanted to do and it was not going to help me much in reconstructing the heart. That judgement was largely correct. Very few models of cell function need to include the individual channels. Stochasticity and averaging out ensure that. This is a problem similar of course to the difference between molecular dynamics and thermodynamics. To understand the properties of a gas in large-scale work, the random motion of the individual molecules does not need to be represented.

Swimming against the tide? In Chapter 5, when describing the painful process of the reinterpretation of i_{K2} as i_f, I used the quotation from Dante's *Purgatorio*, where he makes the great Troubadour, Arnaut Daniel, speak of his pain. Those were not Arnaut's own words of course. But we do in fact have his own words for swimming against the tide in this lovely verse in which he describes his poetic acrobats:

Ieu sui Arnaut qu'amas l'aura	I am Arnaut who gathers up the wind,
E chatz le lebre ab lo bou	And chases the hare with the ox,
E nadi contra suberna	And swims against the torrent.

I first learnt this poem and how to pronounce it from native language speakers in the Périgord, where I have had a village farmhouse for over 50 years. When I first arrived in the village, I had no idea that a language other than French was spoken there. I simply thought that, when I couldn't understand local people, the dialect must be too broad. I had such an experience as a boy when evacuated to Yorkshire during the Second World War – broad Yorkshire dialect spoken by children in the streets was incomprehensible to a London boy. But when I put this idea to a local farmer he bridled: 'Non, monsieur Noble, ce n'est pas notre dialecte, c'est notre langue!' It was clear that if I was to integrate well into the village, I had better learn it. He introduced me to a teacher, Jean Roux (Joan Ros in Occitan), who was deep into studying the Troubadours. He also loved these words of Arnaut Daniel. Most of what I know has been learnt from that family. I write 'family' because in a strange twist of fate, his son, Etienne (Esteve) is also not only a brilliant manipulator of the language (naturally enough) but was also a physiologist working at the University of Bordeaux. We have collaborated (Roux *et al.*, 2001; Marhl *et al.*, 2006; Roux *et al.*, 2006) in extending the physiome modelling approach to the lung. All our email correspondence, even on the science, is in Occitan. When receiving an honorary doctorate from the University of Bordeaux, I was delighted to be able to use the language in my reply.

8.3.2. *Philosophical thread*

The second strand in the transition was philosophical. I wrote in Chapter 2 (p. 22) that a major attraction for me when I left UCL was the extraordinary reputation of Oxford philosophy. It wasn't long before the naivety started to wear off and become replaced with a more professional ... well, at least a veneer of professionalism.

The first interaction was with Dick Hare, the author of *The Language of Morals* (Hare, 1963) and *Freedom and Reason* (Hare, 1965), with whom I discussed the nature of pain. I didn't realise at first that he had been a prisoner of war in a Japanese camp from the fall of Singapore in 1942 right through to the end of the war in 1945. Not surprisingly, his interest in pain and in the way in which philosophy could address the question of the harshest conditions in which humans might live brought an immediacy and urgency to what he wrote and talked about. That itself was fascinating enough to a young physiologist – recall that I still had a neural section in my laboratory in those days. But also of great interest was that, in *Freedom and Reason*, Hare approached ethics almost like a scientific experiment. A moral statement can be compared to a conjecture (hypothesis) which might be refuted if it failed various tests (in particular, the test of universality – could it apply regardless of the individual to whom the moral statement was applied?). In effect, he introduced a version of Karl Popper's principle of falsifiability. I was sufficiently attracted to this idea that I used it once in a sermon in the Balliol College Chapel – successive chaplains in the College have been broad-minded enough not to insist on a strict interpretation of what a sermon should be. Years later, I even gave a sermon from a non-theist Buddhist perspective (Noble, 2008b). Readers of Chapter 10 of *The Music of Life* will know what that was about. All this was important enough, but even more relevant was learning some of the important distinctions from Hare, such as when a movement is an action. Moral philosophy requires that, of course, but so does analysis of the physiological basis of behaviour.

Which leads me naturally on to the next interactions, which were with Anthony Kenny, who has written far more philosophical books than I could even list here, and Alan Montefiore, a philosopher with strong connections to contemporary French philosophy. They both introduced me to the work of Charles Taylor, *Explanation of Behaviour* (Taylor, 1964). There were several important outcomes of those interactions. Taylor and

I debated some of them in articles in *Analysis* (Noble, 1967a, 1967b). Initially, Tony Kenny and I thought that my article was a straightforward knockout of Taylor's ideas – I can't now recall whether Alan agreed with that. When Taylor replied, however, Tony Kenny immediately commented that he had not anticipated how good the reply might be. This introduced me to a fundamental concept in explanation. The knockout nature of my attack was based on what is the case when one considers a single occurrence of an event (specifically in this case a behaviour) where it was easy for me to show that for any higher-level (e.g. teleological) explanation there must always be a difference at the lower level which could also count as the explanation of the event. Taylor responded with the idea that there could be order at the higher level, which could form the basis of an explanation. By contrast, while there would be individual differences at the lower level in each case, they might not, as a set, conform to an explanatory order. Those who are familiar with my argument between a physiologist and a philosopher in Chapter 9 of *The Music of Life* will recognise that this is a distant origin of one of the main ideas in that argument. I am now much closer to Taylor's position than I originally thought, though I did also acknowledge this in the second *Analysis* article. Some of these ideas also found their way into Tony Kenny's book *The Five Ways* (Kenny, 1969). These philosophical interactions are clearly the basis of the realisation that order may exist at one level of a biological system but not at lower levels, even though the higher-level states must have corresponding differences at the lower level. The important insight is that those differences may be strongly variable from one instance to another. Significantly, we now know that this is precisely the way in which different species of birds have adapted to altitudes. The different haemoglobins have the same function, to capture oxygen more easily, but the genetic changes by which this is achieved appear random.

Important though these interactions were, the really sustained formative interaction was with Alan Montefiore. As I have already indicated, Alan straddles the Anglo-Saxon and French worlds of philosophy. We were a natural pair for discussions since we share the fascination with French culture and language. We followed on from the debate with Charles Taylor by running for some years a graduate class in the philosophy of behaviour, on which we were joined by the Psychology/Zoology Fellow of the College, David McFarland, and another philosopher, Kathy Wilkes. Balliol College became a kind of multi-disciplinary centre for the years while those seminars were running. They were successful enough to lead to a book that Alan and I edited (Montefiore and Noble, 1989b) and to which I contributed several chapters (Montefiore and Noble, 1989a; Noble, 1989a, 1989b). Some of this work was also written up for another book (Noble, 1990). This was the context in which I developed the central story of the brain in Chapter 9 of *The Music of Life*. To the insight developed in the debate with Taylor, these discussions added the realisation that many higher-level 'events' are best not described as 'events'. In this sense, an intention is not an event for which we need to look for a particular corresponding neurophysiological event.[2] I will leave the development of this thread to Chapter 9 where I describe the immediate formative elements that led to writing *The Music of Life*. Here, I will just note that it can take years, even decades, to allow a set of philosophical ideas to mature. While the discussions with Alan Montefiore were essential to what eventually led to *The Music of Life*, I could not have written that book at the time that *Goals, No Goals and Own Goals* was written. And, almost certainly, I could not have written *The Music of Life* without those extensive interactions with professional philosophers.

The spirit of multi-disciplinarity still lives on in Balliol. There is now a Balliol Interdisciplinary Institute (BII) that employs modest funding to sponsor graduate student projects.

8.4. Passing on the Baton

I retired from my Oxford chair in 2004. The transition to an Emeritus Professor was relatively painless for a very good reason. Over the preceding years, I had already passed the baton of experimental work onto the scientist who inherited my laboratories.

I first met Peter Kohl (Fig. 8.2) at a conference in Prague around 1992.

He had been working in East Berlin and in Moscow, and his work, at that time, was not well known in the West. His earliest publications were in Russian. Nevertheless, and with restricted facilities, he was carrying out some very ambitious experiments on a controversial area of

---

[2] In those debates, I used a computer analogy to express this insight. A behavioural event could correspond to the implementation at a particular time of a particular instruction. The installation of the program could correspond to the existence of an intention. An intention is then more like a capability, something that is not pinned down to a particular time, just as a computer program gives the machine its capabilities for as long as the program is installed.

Figure 8.2. Denis Noble (left) and Peter Kohl enjoying a dessert wine at Denis's farmhouse in the Périgord. Neither can now recall what the conversation was about, but this photograph formed part of the inspiration for the story of 'pointing behaviour' in Chapter 9 of *The Music of Life*, although the context of the story was changed.

integrative cardiac biology, and I was so impressed with what he showed me that I invited him to come to Oxford as a postdoctoral fellow. That confidence in him was fully justified. His innovative suggestions of myocyte–nonmyocyte electrical coupling and of mechanosensitivity of the electrical activity of fibroblasts were eventually proven by his investigations that went very much against the then-established views of cardiac tissue structure and function. He brought the area of mechano-electric feedback in the heart to Oxford. There had been no previous work of this kind in the department. He was therefore an independent researcher with original projects even before joining my group in the department at Oxford. That innovative approach has continued as he has developed his own independent research team to produce some lovely results. Many people talk about the *potential* of the systems biological approach; Peter *produces* it.

By 2004, Peter was effectively already running the experimental research labs. Handing responsibility over to him was therefore simply a formality, and it also enabled me to continue to interact closely with experimental research while pursuing my work on developing the intellectual base of the systems approach. The only problem we faced was how to manage without the automatic support that came from holding a British Heart Foundation chair. Those funds were critical in enabling me over the

20-year period for which I held the chair to finance new initiatives as pilot projects before they were mature enough to justify full grant applications. Somehow, though, we managed the transition and kept the grant money flowing in. There are now strong experimental and computational cardiac teams both in the department of physiology, anatomy and genetics and in the computing laboratory. Some of the key papers published since or just before my retirement have been in collaboration with Peter Kohl and his team (Kohl *et al.*, 2000; Garny *et al.*, 2005; Iribe *et al.*, 2006; Garny *et al.*, 2009; Kohl and Noble, 2009; Kohl *et al.*, 2010; Rodriguez *et al.*, 2010).

More recently, Peter has also brought his own critical eye to help in the clarification of what systems biology is about (Kohl and Noble, 2009; Kohl *et al.*, 2010).

8.5. Conclusions

8.5.1. *Pluses*

Tracing the threads of a philosophical journey is a major challenge. In these chapters, I have woven together the professional and the personal in order to reveal at least part of what makes a scientist like me 'work'. It is a story with apparently conflicting themes. On the one hand, a 65-year journey along a single path towards understanding the heart, actually just a small part of it: the electrical signals that trigger the heartbeat. On the other hand, a multiple maze rambling through the gardens of philosophy, music, languages, cuisine and much else, yet in an integrated way that leads naturally to enable a book like *The Music of Life* to be written.

From the oriental tradition, I have encountered two poetic inspirations for these contrasting journeys. The first was a favourite of the distinguished Japanese cardiac electrophysiologist, Hiroshi Irisawa, who liked to quote the poem of Ikkyū Sōjun (1394–1481): 'To each fisherman, just one rod',[3] clearly the motto for a single-minded highly focussed search.

---

[3] In Japanese, 漁父生涯竹一竿, or 漁夫生涯竹一竿 (*gyo-fu shougai take ikkan*). This poem is attributed to the great Buddhist monk, Ikkyū Sōjun 一休宗純, who was also the subject of my sermon to the Balliol College Chapel, 'Meditation on two heretics' (Noble, 2008b). He was one of the two 'heretics' in that sermon, having broken several of the central rules of Buddhism, yet he became the highly respected Abbot of Daitokuji, the temple of 'great virtue' in Kyoto, which he restored. The glory of what he restored can be appreciated by wandering and meditating, as I have done, through the gardens and buildings of that beautiful temple.

The contrasting poem comes from a much earlier Korean monk, Won Hyo 元曉 (원효), 617–686, who can be regarded as a predecessor of Ikkyū since he was a comparable heretic, breaking almost the same rules as he did. He wrote, 'Never follow one discipline, but never neglect any discipline', which might be the motto for cross-disciplinary work.

8.5.2. *Minuses*

The dangers of autobiography are well known. Recollected stories are riddled with gaps, and worse, through the vicissitudes of memory. I have tried to avoid the worst errors of this kind by plaguing my colleagues with questions to check my memory. Embellishment and poetic licence also creep in. So much so as time goes by that, eventually, it is almost impossible to disentangle the story from its embellishments. If you think you can do so, remember the famous 1950 Japanese film *Rashomon* 羅生門 by Akira Kurosawa 黒澤明. Rashomon was a gate of the city of Kyoto, ruined after a war. A woodcutter and a priest take refuge in the gatehouse from a downpour. They and other witnesses then relate the story of a rape and murder. The stories conflict; completely so. Yet, each of the witnesses saw the same events. Each account creates distance from the 'truth' by weaving in various justifications. It is one of the most powerful movies I have seen and I still keep a DVD copy to view it from time to time. If you think that only what you see with your own eyes is the truth, you need to watch this film.

8.5.3. *Contribution to systems biology*

Without this journey, I would not have been ready to embrace the ideas of systems biology and to try to develop them. I will take this theme up more completely in Chapters 9 and 10.

References

Bassingthwaighte, J. B., Hunter, P. J. and Noble, D. (2009) 'The cardiac physiome: Perspectives for the future', *Experimental Physiology*, 94, pp. 597–605.

Elwes, R. H. M. (1951) *The Chief Works of Benedict de Spinoza*. New York: Dover.

Fink, M. and Noble, D. (2009) 'Markov models for ion channels - versatility vs. identifiability and speed', *Philosophical Transactions of the Royal Society Series A*, 367, pp. 2161–2179.

Fink, M. and Noble, D. (2010) 'Pharmacodynamic effects in the cardiovascular system: The modeller's view', *Basic & Clinical Pharmacology & Toxicology*, 106, pp. 243–249.

Garny, A., Noble, D., Hunter, P. J. and Kohl, P. (2009) 'Cellular Open Resource (COR): Current status and future directions', *Philosophical Transactions of the Royal Society A*, 367, pp. 1885–1905.

Garny, A., Noble, D. and Kohl, P. (2005) 'Dimensionality in cardiac modelling', *Progress in Biophysics and Molecular Biology*, 87, pp. 47–66.

Hampshire, S. (1956) *Spinoza*. London: Faber & Faber.

Hare, R. M. (1963) *The Language of Morals*. Oxford: Oxford University Press.

Hare, R. M. (1965) *Freedom and Reason*. Oxford: Oxford University Press.

Hunter, P. J., Li, W., McCulloch, A. and Noble, D. (2006) 'Multi-scale modeling: Standards, tools and databases for the Physiome Project', *Proceedings IEEE*, 39, pp. 48–54.

Iribe, G., Kohl, P. and Noble, D. (2006) 'Modulatory effect of calmodulin-dependent kinase II (CaMKII) on sarcoplasmic reticulum Ca^{2+} handling and interval-force relations: A modelling study', *Philosophical Transactions of the Royal Society A*, 364, pp. 1107–1133.

Jack, J. J. B., Noble, D. and Tsien, R. W. (1975) *Electric Current Flow in Excitable Cells*. Oxford: Oxford University Press.

Kenny, A. J. P. (1969) *The Five Ways*. London: Routledge & Kegan Paul.

Kohl, P., Crampin, E., Quinn, T. A. and Noble, D. (2010) 'Systems biology: An approach', *Clinical Pharmacology and Therapeutics*, 88, pp. 25–33.

Kohl, P. and Noble, D. (2009) 'Systems biology and the virtual physiological human', *Molecular Systems Biology*, 5, 292, pp. 291–296.

Kohl, P., Noble, D., Winslow, R. and Hunter, P. J. (2000) 'Computational modelling of biological systems: Tools and visions', *Philosophical Transactions of the Royal Society A*, 358, pp. 579–610.

Marhl, M., Noble, D. and Roux, E. (2006) 'Modeling of molecular and cellular mechanisms involved in Ca^{2+} signal encoding in airway myocytes', *Cell Biochem Biophys*, 46, pp. 285–302.

Montefiore, A. C. R. G. and Noble, D. (1989a) 'General introduction', In *Goals, No Goals and Own Goals* (ed. Montefiore, A. C. R. G. and Noble, D.), pp. 3–13. London: Unwin-Hyman.

Montefiore, A. C. R. G. and Noble, D. (ed.) (1989b) *Goals, No Goals and Own Goals*. London: Unwin-Hyman.

Niederer, S. A., Fink, M., Noble, D. and Smith, N. P. (2009) 'A meta analysis of cardiac electrophysiology computational models', *Experimental Physiology*, 94, pp. 486–495.

Noble, D. (1967a) 'Charles Taylor on teleological explanation', *Analysis*, 27, pp. 96–103.

Noble, D. (1967b) 'The conceptualist view of teleology', *Analysis*, 28, pp. 62–63.

Noble, D. (1989a) 'Intentional action and physiology', In *Goals, No Goals and Own Goals* (ed. Montefiore, A. C. R. G. and Noble, D.), pp. 81–100. London: Unwin-Hyman.

Noble, D. (1989b) 'What do intentions do?' In *Goals, No Goals and Own Goals* (ed. Montefiore, A. C. R. G. and Noble, D.), pp. 262–279. London: Unwin-Hyman.

Noble, D. (1990) 'Biological explanation and intentional behaviour', In *Modelling the Mind* (ed. Said, K. A. M., Newton-Smith, W. H., Viale, R. and Wilkes, K.), pp. 97–112. Oxford: Oxford University Press.

Noble, D. (2007) 'From the Hodgkin-Huxley axon to the virtual heart', *Journal of Physiology*, 580, pp. 15–22.

Noble, D. (2008a) 'Computational models of the heart and their use in assessing the actions of drugs', *Journal of Pharmacological Sciences*, 107, pp. 107–117.

Noble, D. (2008b) 'Meditation on two 'heretics'', In *The Balliol Record*, pp. 28–30.

Rodriguez, B., Burrage, K., Gavaghan, D., Grau, V., Kohl, P. and Noble, D. (2010) 'Cardiac applications of the systems biology approach to drug development', *Clinical Pharmacology and Therapeutics*, 88, pp. 130–134.

Roux, E., Noble, P. J., Hyvelin, J.-M. and Noble, D. (2001) 'Modelling of Ca^{2+}-activated chloride current in tracheal smooth muscle cells', *Acta Biotheoretica*, 49, pp. 291–300.

Roux, E., Noble, P. J., Noble, D. and Marhl, M. (2006) 'Modelling of calcium handling in airway myocytes', *Progress in Biophysics and Molecular Biology*, 90, pp. 64–87.

Stewart, P., Aslanidi, O. V., Noble, D., Noble, P. J., Boyett, M. R. and Zhang, H. (2009) 'Mathematical models of the electrical action potential of Purkinje fibre cells', *Philosophical Transactions of the Royal Society Series A*, 367, pp. 2225–2255.

Taylor, C. (1964) *The Explanation of Behaviour*. London: Routledge & Kegan Paul.

Ten Tusscher, K. H. W. J., Noble, D., *et al.* (2004) 'A model of the human ventricular myocyte', *American Journal of Physiology*, 286(4), pp. 1573–1589.

Chapter 9

Systems Biology and *The Music of Life*

9.1. Basel and Music 2003

In 2003, I was invited to lecture on systems biology of the heart in Basel at a meeting organised by Torsten Schwede at the Biozentrum of the University of Basel. The evening before the lecture, a speakers' dinner was held in one of the ancient cellars of the city. A classical guitar recital preceded the dinner. I suspect that the organisers knew my love of the guitar. They had invited a world-ranking performer, Christoph Denoth, not only to perform but also to join the dinner afterwards, with a place allocated next to mine. The performance was inspired. Christoph is to music what some of my brilliant colleagues have been to mathematics: someone who moves so easily in the medium that they can focus on inter-pretation and inventiveness rather than just on technicality. One of the pieces was a *homage* that Christoph himself had composed. He rarely plays it, but that evening, the tremolo parts of the piece were magical. He made the guitar perform successively all the roles of a balalaika, a multiple-harmonic percussion instrument, and a sweet lute in the same piece of music. No other instrument can match the classical guitar in the hands of a world-class performer.

It is easy therefore to imagine my disappointment when, at the end of the recital, he started walking out of the room. Not knowing whether this was shyness – perhaps not relishing dining with a bunch of academics – I reacted instinctively, rushed over to intercept him and asked whether I could see his guitar. I couldn't think of any other way to detain him. He took it out of its case and, as he handed it to me, he noticed the difference in the lengths of the fingernails on my two hands. 'You play!' he exclaimed (the left-hand nails are short to enable the fingers to firmly stop the strings at the various positions on the fingerboard, while those of the right are carefully trimmed to be at the optimal shape and length for playing the strings). 'Yes, but not like you!' I replied. Before I knew how to avoid it, he had struck a deal: 'I join the dinner if you agree to play a piece after the main course'. Was he joking?

Just as I wouldn't even presume to compete with my mathematical colleagues who 'see' solutions even before they prove them, I wouldn't dare to follow a performer of Christoph Denoth's international level with an attempt at a classical piece. Thinking quickly, I said that I would play, but I decided to accompany myself simply with a song, rather than playing a classical solo. Arpeggio accompaniment of a song is much safer than attempting to play a Villa-Lobos prelude! And, by then, I had had some years of experience of public performance of my interpretations of love songs in the language of the medieval Troubadours, Occitan (often known in the Anglo-Saxon world as Provencal). So, after the main course, I sang *Arron d'Aimar* (after love), a modern Gascon song in Occitan composed by one of the most successful groups in the Pyrenees, *Nadau*.

Someone who speaks French or Italian would follow a few of the words of a song in Occitan. It is not hard to see that 'que t'aimi' is 'que je t'aime' (how I love you) even when it is pronounced quite differently (in Occitan it sounds like 'kay tie me'). Christoph was not, however, catching just the odd phrase. He was nodding all the way through. As a remarkable coincidence, he is one of the small minority (roughly 30,000) of Swiss people who speak Romansch, a language that has similar roots to Occitan. We spent the rest of the evening comparing notes on the two languages and what we perform on the classical guitar.

The next day, as I was giving my lecture, I noticed that, up there near the back of the lecture theatre, Christoph Denoth had crept in to listen. I had just reached the point at which I was explaining that it is wrong to think of genes as controlling the organism, that this is to put the cart before the horse, as it were. His presence led me to look for a way of relating this to the playing of music. Almost without thinking (I can often listen to myself giving a lecture when I am doing so in free-style presentation), I said, 'You know, it would be just as absurd to think that the pipes in a large cathedral organ determine what the organist plays. Of course, it was Bach who did that in writing the score, and the organist himself who interprets it. The pipes are his passive instruments until he brings them to life in a pattern that he imposes on them, just as we impose patterns on our genomes to generate all the 200 or so different types of cells in our bodies'.

And so, the title of my book, *The Music of Life*, was born and the message of Chapter 2 ('The organ of 30,000 pipes') was virtually already written. I did not know then that there are pipe organs that have as many pipes as the human genome has genes (roughly 25,000 – see footnote on page 31 of *The Music of Life*). Nor did I know that Barbara McClintock, who won the Nobel Prize late in life (in 1983 at the age of 81 for her discovery of jumping genes), had referred to the genome as an 'organ of the cell' (McClintock, 1984), responding not only to the commands of the cell but also reacting to the challenges of the environment. Yet, here we are, over 40 years later, still receiving new students at our universities who, at school, have absorbed the dogma that 'they [genes] created us, body and mind' (Dawkins, 1976, 2006). *The Selfish Gene* is, often enough, the only book on genetics that they have read, and we as university teachers have to undo the damage. Not only us; Richard Dawkins himself also has had to distance himself from the naïve genetic determinism – a later book (Dawkins, 2003) has a chapter entitled 'Genes aren't us'.

Barbara McClintock worked on plants (mainly maize). The insight that the genome is primarily a database (another way of expressing 'organ of the cell') used by the cells, tissues, organs and systems of the body is just as true for other organisms. As Beurton, Falk and Rheinberger (2008) comment, 'It seems that a cell's enzymes are capable of actively manipulating DNA to do this or that. A genome consists largely of semistable genetic elements that may be rearranged or even moved around in the genome thus modifying the information content of DNA'.

9.2. Oxford 2005 – Gestation of *The Music of Life*

If Basel was its conception, Oxford was its gestation. There were several strands of intellectual activity that combined to form its womb.

9.2.1. *The scientific womb*

The first was my own and my collaborators' scientific research over many years. To say that the 'Selfish Gene theory' was useless to us as a hypothesis would be understating the problem. It has no cashable value whatsoever in physiological work. Indeed, popular neo-Darwinism in general relegates the organism and its development to the category of a transient vehicle – the 'lumbering robot' destined to transmit its 'eternal' genes on to the next generation. All of this is biased metaphorical polemic, and I would argue that it has impeded the systems approach (Noble, 2011b).

Modern evolutionary biologists now work to extend (see Pigliucci & Müller, 2010) the Modern Synthesis (the more correct term for what is often called neo-Darwinism) and to reconnect with development in the field of Evolutionary Developmental Biology (Evo-Devo). But it was not just development that was ignored. Physiology was too. Yet, it is whole organisms that live or die, which is the process essential to selection. The only way in which that could be reinterpreted to be the same as the selection of genes was to maintain, as Dawkins does in *The Extended Phenotype* (Dawkins, 1982), that the genes-eye view and the organism-eye view are essentially equivalent, just perceived in a different way (rather like the Necker Cube illusion used on the front cover of the first edition of *The Extended Phenotype*), so that the one can replace the other. That would be true only if one really could identify each phenotype characteristic with a difference at the level of DNA. As I explained in Chapter 6, that is simply not true. Most of the time, as physiologists, we are bedevilled by the robustness of the organism in resisting the consequences of changes at the genetic level. Moreover, it is incorrect to consider important functions, like the rhythm of the heart, as being determined by 'genetic programs'. There are no such programs. The example I gave in Chapter 6 illustrates this with the necessity of including downward causation from the complete cell as an absolutely essential part of the process of pacemaker activity, such that the attribution of cardiac rhythm simply doesn't make sense at the level of genes and proteins. I have explored this problem in the article on 'Differential and Integral Views of Genetics' (Noble, 2011a).

9.2.2. *The philosophical womb*

The second was academic philosophy. Many of my scientific colleagues would consider the discipline of philosophy to be a waste of time. On this view, science has progressively replaced philosophy in providing answers to important questions about the nature of the world and of ourselves. Clearly, in unravelling facts about the world through the process of empirical discovery, and in formulating conceptual schemes to create mathematical and other formal descriptions of those facts, this is undeniably true. What used to be called 'natural philosophy' is now called science. The remnants of this history can be seen in the title of one of my favourite journals, the *Philosophical Transactions of the Royal Society*, the longest-running scientific journal in the world, and in the titles of some scientific chairs in British universities.[1]

So, can we ignore philosophy, relegating it to the category of things about which we cannot speak (to quote the ending of Wittgenstein's *Tractatus*)? This is about as impossible as it would be to eat ourselves to survive starvation! As soon as we open our mouths to speak, or take up our pens to write, we are guilty (if that is the right word for an action we cannot avoid) of being philosophers: all of us – no exceptions, including those who may loudly proclaim that they are not philosophers; indeed, they are the most likely to err. The reason is simple but subtle. Our language contains conceptual traps for the unwary at every turn of phrase. Conceivably, science might succeed in being entirely neutral, philosophically speaking, if the only language it used was that of mathematics and logic (though even there I am not entirely sure – even in logic we have to choose the form of logic we employ – do we allow what Buddhists call four-cornered logic, for example?) But to restrict science in this way would not be to produce empirical discovery. Mathematics and logic are conceptual tools, not in themselves forms of empirical statements. So, we must also use declaratory language as well.

But as soon as we use language, whether ordinary or technical, the conceptual frame in which our language has developed creeps in, surreptitiously, to colour our thought and expression. How else are we to understand the way in which neuroscientists like Sherrington (1940) and Eccles (1953) were 'trapped' in their dualist interpretation of the relation between the brain and the mind (see Chapter 9 of *The Music of Life*)? The reification of 'self' was a historical linguistic development in Western thought (Hacker, 2011), not a necessary 'object' for neuroscience to discover. Hardly anyone today amongst scientists thinks that the dualism of Sherrington or Eccles was anything other than a philosophical prejudice. This kind of prejudice creates the greatest problems when we are unaware

---

[1]Examples include the Kelvin Chair of Natural Philosophy (Glasgow) and the Chairs of Natural Philosophy in Cambridge and Edinburgh, and, of course, the basic doctorate in our universities is the PhD, Doctor of Philosophy.

of the surreptitious nature of the philosophical baggage we import into our 'science'.

Here, I will take just one example of the problem. What 'causes' an organism? What, indeed, is life, to take the title of Schrödinger's famous book (Schrödinger, 1944)? I had, briefly, thought of using the same title for *The Music of Life*. I didn't dare to do that, but I was concerned that the question had been badly misunderstood, which is why the Introduction of my book starts with the question.

Take, for example, Francis Crick's statement:

> You, your joys and your sorrows, your memories and your ambitions, your sense of personal identity and free will, are in fact no more than the behaviour of a vast assembly of nerve cells and their associated molecules. (Crick, 1994)

It is difficult to know how to take such a statement. Is he joking? Crick was famous for his sense of humour and fun. But this statement essentially is *The Astonishing Hypothesis*, the title of the book. So, it can't just be dismissed as a humorous aside. It is based on a philosophical position, which is that the lower elements in a system (in this case, the cells and molecules) are always to be preferred in seeking an explanation for the higher elements (memories, ambition, identity and freewill – the latter two only as the 'sense of', which is already to imply that, somehow, they are just what we perceive, so not real). This is a prejudice, though a common one in scientific writers. There is absolutely no reason, a priori, to favour one level over another in seeking the 'causes' of anything. To anticipate my next section, 'level', 'higher' and 'lower' are metaphors, though not always recognised as such.

Let's spell this out in terms of one of the examples I considered in Chapter 6. When we construct a model of cardiac rhythm, we do so by creating differential equations for each of the 'lower' elements in the cell that is involved, i.e. the proteins forming channels, transporters or buffers. To these we add equations for the structure (cell, organelles). Then, we solve the equations by integrating them. At that point, we necessarily incorporate initial and boundary conditions without which the integration would be impossible. If we look for the causes of the solution, these conditions are just as much a cause as the differential equations themselves. It wouldn't make sense to say that the rhythm is 'nothing more than the activity of the molecules' precisely because the physical structure *is* something more. It constrains what the molecules do, even to the extent of rhythm being abolished if we remove that constraint. As I wrote in Chapter 1, it doesn't even make sense to ascribe cardiac rhythm to the molecules of the system.

Philosophers have analysed the concept of cause over millennia, starting at least with the work of Aristotle. It is an elementary aspect of work in philosophy to recognise the different categories of causation. I explore this kind of question in relation to genetic causation in one of the articles (Noble, 2008) from *Philosophical Transactions of the Royal Society*, 'Genes and Causation', which seeks answers to the following questions:

(a) How has the concept of 'gene' changed since its invention 100 years ago and how does that change how we view genetic causes? The answer is that the change is fundamental. From being a necessary cause (defined in that way as a hypothetical entity, postulated as an allele – a gene variant), a gene has become a particular physical, far-from-hypothetical, entity for which causation has to be demonstrated experimentally. Each DNA sequence that we identify as a gene can have causal consequences far outside the phenotypic domain under which it was first named.

(b) If the genome inheritance can be represented as digital information, how do we compare it with the analogue information inherited with the fertilised egg cell? I argue that the non-DNA information is at least as important. In terms of information, they are comparable (Noble, 2011a).

(c) Does it make sense to view one side of this duality of inheritance as primary? My answer is that there is no such reason. They are different, but both are completely necessary. If we wish to give primacy, though, I would favour the cell. That is why viruses are 'dead' outside the context of a cell. If you enucleate a cell, it continues to function (red cells in the blood do just that) until it needs to make more protein. But if you destroy the cell membrane, you no longer have functionality.

'Genes and Causation' also introduces what I call the 'genetic differential effect problem', which is produced by the neo-Darwinist insistence that what matters is differences at the genetic level that cause differences at the phenotype level. This issue is pursued at greater length in Noble (2011a).

9.2.3. *The metaphorical womb*

The third form of gestation was the theory of metaphor. At the time that I started working on *The Music of Life*, I was advising a Korean colleague, Sung Hee Kim, actually the wife of Yung Earm (see Chapter 5), on some of the medical terms used in body metaphors since she was working on a doctoral thesis in the Faculty of English Language and Literature of Oxford University comparing body metaphors used in Korean and English print media (Kim, 2006). Her research identified some major differences between the two languages that could only be understood by reference to the full semantic frame of the metaphor in each language and which could be best represented by mapping the metaphorical statements onto their targets in a systematic way. A key aspect of metaphor theory used in this kind of approach is that competing metaphors illuminate different parts of the target to which they are applied. By doing so, they can both be correct. As a simple example from the work of metaphor theorists like Lakoff & Johnson (1980), love can be a journey and it can be war. The two metaphors illuminate such different aspects of the same target, love, that – as we all know only too well – they can both be true.

The trap here is that science likes to avoid competing explanations that can both be correct. We like to isolate a problem from all the complicating factors until we can see an experiment that unambiguously decides between the possible theories to arrive at a single 'true' answer. But if we have imported metaphorical ideas into our theories, how can we be sure that this is the correct approach? Competing metaphors can both be correct.

Scientists are sometimes not aware, first, that they use metaphors much more frequently than they might think and, second, that the relationship between a metaphor and reality is very different from that between a scientific theory and reality. This is the issue I explore in Chapter 1 of *The Music of Life* by comparing 'selfish' and 'prisoner' genes to demonstrate their metaphorical nature. Just like 'journey' and 'war' applied to 'love', there is no experiment that could unambiguously disprove their application. They can both be 'true'. In short, they are not empirical scientific theories. 'The Selfish Gene' is an idea more in the field of metaphorical polemic than science (Noble, 2011b; Noble and Noble, 2020).

Could 'The Selfish Gene' idea be rescued for science? It has been so phenomenally successful as a popularisation of genetic biology that it would be valuable if it could be interpreted as a standard scientific hypothesis. At least we would then be able to subject it to the standard scientific test of validity and see how and whether it could be falsified. I can see only one way in which that could be done. It would require a criterion of 'selfishness' that could be assessed at the level of genes (now defined as particular DNA sequences) *independently of the prediction that the theory makes*, i.e. that the frequency of that gene in the gene pool increases in subsequent generations. Without such a criterion, the idea is circular, as metaphorical ideas often are.

Try as one might, though, the task of finding such a criterion is surely and inevitably a failure (Noble, 2011b). How could a particular nucleotide sequence, the only measure we have at the level of genes, tell us whether that gene is or is not selfish? The only test we have is the prediction of the statement itself, i.e. that the frequency of that gene increases in the gene pool – that is what is *meant* by 'selfish' here. It is a strange 'theory' that depends on its own, *and only*, prediction even to have meaning, let alone be testable. Moreover, that meaning depends on the context, i.e. the phenotype and the environment. A sequence that may be successful in one organism in a particular environment could be disastrous in another organism or in a different environment. And, indeed, virtually all cross-species clones fail to survive. The DNA has been put into the wrong context. The sequence by itself means nothing, just as a word of Maori might have no meaning or, worse still, the wrong meaning in English. In looking for an independent criterion to save the theory from circularity, we are therefore forced to move to the level of the phenotype after all. The concept of the selfish gene is then redundant and misleading. That, in a nutshell, is why the concept is of no use in physiological science (Noble, 2011b).

9.2.4. *The historical/religious womb*

I don't think of myself as a religious person, though I hold on to the idea of human spirituality, which is itself a 'high'-level concept not capable of molecular- or cellular-level reduction[2]. I would also argue strongly that the

---

[2] 'Spirit' comes from the Latin for breath, *spiritus*, and it is not incidental that many meditation techniques focus on the breath. 'Respiration' comes from the same root. So does 'inspire'. There are many modern definitions of 'spirit'. Here, I am using the term as almost interchangeable with, but also as an extension of, the 'self', which I think should be viewed as a process, not an object. Processes can be causes. Viewing 'spirit' or 'self' in

idea of spirituality does not depend on particular metaphysics. In fact, my stance is basically non-metaphysical, as will be clear from the last two chapters of *The Music of Life*, which is what brings me closer to the Buddhist tradition than to any other. I see questions to which the answers are largely a matter of personal choice amongst rival metaphysics, as beyond what empirical science can tell us. But that applies both ways, to theism and atheism alike. So, I don't find the high-profile (and usually very naïve) attacks on religion by some scientists to be either convincing or helpful. They fail to address the central problem, which is that metaphysical questions are notoriously difficult to express coherently. Many of the questions are incoherent because we simply do not know what it would be for them to have an answer. Yet, as human beings, we find it difficult (theists, atheists and agnostics alike) to avoid the questions. On the one hand, it seems sensible to deal only with what we can observe, measure and understand. This is the pragmatic approach of science. Every valid scientific theory should be falsifiable, at least in principle.[3] On the other hand, it is laughably presumptuous to suppose that this resolves all questions about life. Clearly, it can't. Thus, it is relevant to our ethics in relation to human life to understand genetics, evolution and physiology – we need to know what we are made of and how we developed since those studies set limits to what is possible – but those subjects do not dictate to us what our ethics should be. Attempts to develop a purely scientific humanism as a substitute for religion don't succeed for precisely this reason.

Perhaps a major part of the problem is that, just as we have unnecessarily reified 'self' and 'spirit', we have a limited conception of what

---

this way does not detract from causal efficacy. Someone who denies human spirituality would, I think, be denying such efficacy, relegating whatever is being referred to as an epiphenomenon. I think that position is incoherent. I needed 'inspiration' in the relevant sense to write this book. That is not just a function of my own brain. It is interpersonal. Spirituality therefore has social connotations that take us outside the range of biology. The boundaries of spirituality cannot therefore be precisely defined in biological terms.

[3] Here I am using 'theory' in the sense of an empirically testable proposition about the world. 'Theories' that are not testable in this way, such as mathematics and metaphysics, are better viewed as tautologies that help us in viewing and understanding the world. No experiment could ever disprove a valid geometry, for example. Nor can a 'theory' like 'force = mass × acceleration' be disproved since it has in effect become a definition of force or mass (take your choice). Some biological 'theories' are also in this category. As discussed earlier in this chapter, no experiment could ever disprove (or prove) the 'selfish gene theory', which is therefore also metaphysical.

'religion' means. I was therefore very intrigued indeed when my early attempts to learn East Asian languages (Japanese, Korean and Chinese) led me to an important discovery. The words for 'god', 'religion' and 'prayer' simply do not correspond precisely to the meanings we give them in the Western religious traditions. The 'god' word, 神 (*kami* in Japanese, *shén* in Chinese and *shin* in Korean), carries meanings more like 'spirit' or 'essence'. This is why Christian missionaries to East Asia had to use different Chinese characters to express the concept of a Creator. The word for 'religion', 教, carries more the meaning of 'teaching'. It is used in the compound for 'professor', for example. The context in which these words have their meanings can allow for a 'religion' that, by our Western ideas, might not even count as a religion. Taoism and Buddhism seem to me to be in that category. I was therefore delighted to read *The Awakening of the West* (Batchelor, 1994), a brilliant historical account of the encounter between Western missionaries and Eastern religions. The author was a Buddhist monk for many years in both the Tibetan and Korean traditions. Batchelor's later books (Batchelor, 1997, 2010) are also highly relevant, although I did not know about those when I wrote *The Music of Life*.[4]

This was the background to the story of Jupiterians in the last chapter. But what led me to incorporate such ideas into a book on systems biology? The answer is another convergence between my own thinking and some aspects of oriental philosophy. As I came to write Chapter 9 on the brain, I was reflecting on how to apply the systems approach to neuroscience. Some of my previous articles (Egan *et al.*, 1989; Noble, 1989, 1990; Noble and Vincent, 1997; Noble, 2004) had already developed the idea that the self is a construct, a useful one of course, but not one to be identified either with an immaterial substance or simply with the brain. The way I express this in Chapter 9 of *The Music of Life* is to say that it is better regarded as a process than as an object. Just as it doesn't make sense to talk about heart rhythm at the level of genes and proteins, it doesn't make sense to talk of the self at the level of neurons or hormones. At those levels, it is as though there is no self at all. The idea of no-self (*anatman* in Sanskrit, *an* = no, *atman* = self) is, of course, precisely that of Buddhism.

Or is it? It has taken me several years to try to answer that question. The original insight 2500 years ago may have been part of the general

---

[4] Interested readers can view an extensive discussion between Stephen Batchelor and Denis Noble on http://www.voicesfromoxford.org/B-S-Batchelor.html.

non-metaphysical stance of the historical Buddha, Siddhartha Gautama (Batchelor, 2010), but it is hard to decide precisely what this insight was. We live in such a different world from that of Gautama and it is all too easy, as Gombrich (2009) warned, to take his words out of context. I started out thinking that it was an empirical discovery. Perhaps, during meditation, he looked for the self, the 'I', the soul, and simply didn't find it, rather as David Hume famously examined his thoughts and perceptions two millennia later. I came to the conclusion that none of them could be identified as 'the self', in the sense such a thing did not exist.

But, to say that something doesn't exist, we do at least need to know what it would mean for it to exist, how we would recognise it if we tried to find it. And, of course, we don't know how to recognise it. I recognise you, the reader, as a person, as having a sense of self, and we know what words like 'yourself', 'myself' and 'himself' mean. To indicate these, we would point at you, me or him as the case may be. You can also point at yourself to indicate yourself. But, if we had your brain out on a dish, as it were, how could we possibly say that this is you? The brain is necessary to you, but it is not sufficient. That is the basis of my story of 'the frozen brain' in Chapter 9 of *The Music of Life*.

Looking at the question this way, we are forced to say that the concept of no-self is, just that, a conceptual truth, not an empirical one. No scientific or meditative experiment is necessary to establish such a truth. To return to the quotation from Crick earlier in this chapter, looking for such things at the level of neurons and molecules is a conceptual mistake.

In Chapter 10 of *The Music of Life*, I used the famous ox herder parable (Wada, 2002) from the Chinese Buddhist tradition as a way of explaining the object of meditation to, as it were, subdue the self. One of the ten pictures is just of an empty circle, as though the self (the ox in the story) has disappeared. I no longer think of it this way. It is rather a parable about how to subdue *selfishness*, not the self. Buddhist meditation has, as one of its aims, to remove selfish, greedy and angry attitudes, one of the central aims of any ethical practice.

So, there are two kinds of 'discovery' here. The first is the conceptual truth that it doesn't make sense to talk of the self as an object in the sense in which our brains are objects. The second is that, through meditative techniques, we can subdue selfishness. But doing that is not equivalent to some conjuring trick of 'making the self disappear'. I am reinforced in that conviction by the idea that what the Buddha was arguing against was not so much the self, as usually conceived when we refer to 'himself' or 'myself', but rather against the idea that it was an unchanging thing (Gombrich, 2009). That idea fits well with the concept of the self as a process, as Gombrich also argues.

Does that mean that our experience, e.g. of meditation, is irrelevant? I don't think so. Experience can lead us to a conceptual truth even when it is not itself necessary to that truth. It was seeing the images of gravitational lensing produced by the Hubble telescope that led me to take the idea of the bending of space by huge gravitational fields seriously. Yet, the theory of general relativity does not require me to have that experience in order for it to be a valid theory of the structure of the universe.

I hesitated in writing Chapters 9 and 10 of *The Music of Life*. They were the most difficult to write. The book could have finished on evolution in Chapter 8. But that would have cut its head off. You can't ask a question as audacious as 'what is life' and not deal with questions of the brain and the self.

9.3. *The Music of Life* – Reactions

It has been 19 years since *The Music of Life* was published. It has already been translated into twelve other languages.[5] While being carefully written (don't be misled by the easy, superficially simple style), it does not

---

[5]The translations have been a remarkable experience. My former collaborator Carlos Ojeda (see chapter 5) produced the first, in French, in collaboration with a linguist, Véronique Assadas. He had been involved in many of the debates since the beginning, including a debate at Holywell Manor with Richard Dawkins in the year 1976, when *The Selfish Gene* was published. The philosopher Anthony Kenny (see Chapter 8) challenged the concept of the selfish gene by noting that by knowing the alphabet one did not thereby know Shakespeare, to which Dawkins's reply was 'well I am just a scientist, I am only interested in truth'. From the back of the room, Carlos rapidly interjected 'and what is truth?' Carlos put his heart and mind into the French translation of *The Music of Life* (*La Musique de la Vie*). Sadly, he died before it appeared. It stands as a tribute to his many intellectual contributions to the debate.

More recently I have had the very different pleasure of interacting for a year with the translators, Zhang Li-Fan and Lu Hong-Bing, working on the Chinese version 生命的乐章. We have worked together through many of the problems in trying to express some of the language of *The Music of Life* in Chinese. The experience fully confirms the difficulties of translating when the semantic frames are so different, to which I drew attention in the book.

hide its agenda, which is a radical revision of the way in which we think about and practice biological science. It turns one biological dogma after another upside down, starting with the central dogma of molecular biology and its misinterpretations. So, what is missing? Quite simply, there is no reply from those who might wish to defend the standard biological story that held sway for most of the 20th century. Perhaps that is because people have misunderstood the implications. *The Music of Life* was deliberately written as a short, accessible book. The background of ideas that formed its basis requires much more explanation and exploration, a project that I regard as an ongoing one as I work with colleagues in Oxford and elsewhere to explore in greater depth what the conceptual foundations of the systems approach might be. Those attempts will form the last chapter, 10, of this book.

The second of the *Philosophical Transactions* articles relevant here, 'Biophysics and Systems Biology', was commissioned as part of the celebrations of the 350th anniversary of the Royal Society. Each of those commissioned to write was asked to provide a state-of-the-art assessment and an indication of where the field is going. The article goes well beyond the previous article, 'Genes and Causation', by explicitly dealing with the status of neo-Darwinism and why it is incompatible with more recent discoveries in the physiological and related sciences.

It is of course hazardous, if not impossible, to foresee where any field of science is going to develop in the future. If we knew the future of discovery, it would no longer be discovery. In responding to the invitation of the Royal Society, I focused instead on where the field of systems biology is now and what the issues are that need resolving.

First, is it a field? Is it a subject that should be a separate department in a university? The argument in my article with Peter Kohl, Edmund Crampin and Alex Quinn (Kohl *et al.*, 2010) is that it is not a separate discipline. It is an *approach* that can be applied to any discipline and to all levels in biology. That issue lies at the heart of the difficulties people have experienced in trying to define it. There is a sense of unease, which is frequently obvious when one attends conferences on systems biology. A *Nature* staff reporter clearly expressed this when, interviewing me after lecturing at such a meeting, she challenged me to say what was different about the meeting compared to many other meetings she had to report on in the fields of biochemistry, molecular biology and even physiology. Wasn't there the same almost endless genomic and proteomic data sifting? Was it really just about analysing the mountain of data?

As I show in some of the quotes from Sydney Brenner in *The Music of Life*, he established some of the ideas that form the systems approach: 'I know one approach that will fail, which is to start with genes, make proteins from them and to try to build things bottom up' (stated at a Novartis Foundation meeting in 2001). He first used the term middle-out to distinguish the approach from purely bottom-up or top-down methods (Brenner *et al.*, 2001). Multilevel analysis is central to the systems approach. More genomics, more information sifting, is not the systems approach.

To explain why, let me use another story of the Silmans, who appear from time to time in *The Music of Life*. In Chapter 7, I let these visitors from space be so tiny that they make the mistake of thinking that the cells in a human body are separate organisms, as indeed they are in a sense – multicellular organisms evolved by single cells coming together in symbiotic relationships. One of the reviewers on Amazon describes this chapter as a 'brilliant description of sexual intercourse'. So it is, and it is quite a different story from most such descriptions, but before you rush out to buy it for that reason, he goes on to write, 'that should utterly dispose of any simplistic ideas of "Lamarckian" inheritance of acquired characteristics as "wrong."' Well, maybe that is even more shocking. It should be. Lamarck has not exactly had good press in the English-speaking world for nearly 200 years. The significance of his work in introducing the concept of biology as a science and in establishing the transformation of species is largely ignored, while he is incorrectly represented as the inventor of 'Lamarckism'. In that sense, Darwin was just as much a Lamarckist (Noble, 2010). We need to reassess both Lamarck and Lamarckism. The unjustified denigration of Lamarck is one of the great historical and philosophical mistakes of neo-Darwinism. As Steven J Gould noted, this denigration began at the time of his death when Cuvier (who bitterly opposed Lamarck's arguments for evolution) wrote, 'one of the most deprecatory and chillingly partisan biographies I have ever read'. Metaphorical polemics, masquerading as science, are dangerous to reputations!

But to return to the Silmans, the story of Chapter 7 was based on their being too small to appreciate the significance of a multicellular organism, and so they interpreted each cell as a separate organism and the cell types as separate species. But, through this misunderstanding, they came to appreciate an important fact about how the same genome can be used to make such fundamentally different cells as bone cells and heart cells. Faced with the challenge of multilevel systems biology, we are in a similar situation, not because of our physical size, but rather because of our intellectual immaturity. It is easy to state the multilevel principle of systems biology and to appreciate the idea of no privileged level of causality. But, at the present time, we do not possess the mathematical tools to achieve this in anything other than a piecemeal fashion. As described in Chapter 8, we use an intuitive feel for the significance of events at one level to decide what details to carry up into higher levels. We are a little like the mice in the cartoon, each of which tries to identify what the elephant is.

The cartoon in Figure 9.1 represents systems biology as having the overall view, enabling it to see the whole of life. But it also represents it as detached from the rest, as though it lacks the tools to implement its vision of the whole.

What should those tools be? I think we face a situation not dissimilar to the one I described in Chapter 3, when my focus turned to analytical approaches with closed-form solutions. The problem then was inadequate computing power. Even though we now have massive computing power compared to 65 years ago, we still don't have enough, and if digital computing is the only form we can use, then I think we will never have enough.

Are we simply going to continue developing ever more complex differential equation models to demand ever larger computing power? There are two reasons at least to doubt that.

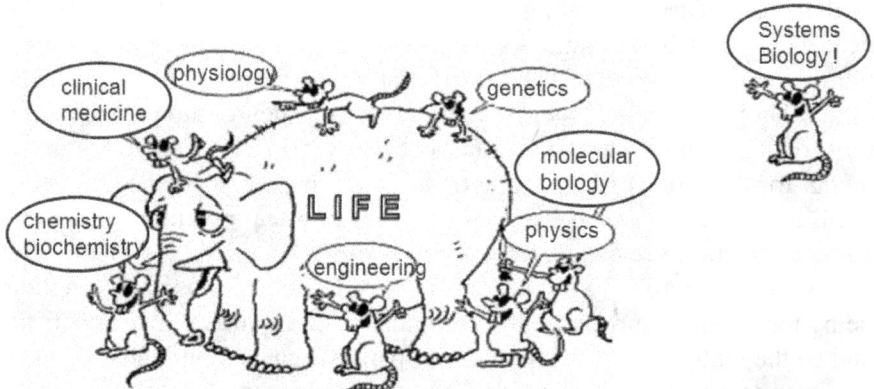

Figure 9.1.   Cartoon produced by Yung Earm to introduce a lecture by Denis Noble at the IUPS World Congress in Kyoto, Japan, in July 2009. The cartoon was inspired by the ideas of the Korean Buddhist monk, Won Hyo (617–686).

First, simply building ever more complex models doesn't necessarily lead to greater understanding, since the models may be sufficiently complex to be, in themselves, in need of explanation. As I showed in earlier chapters, analytical mathematics gives greater generality and greater understanding. The kinds of mathematics I described there involved obtaining closed-form solutions. But that is not the only form of mathematical analysis possible.

What kind of mathematics is required then? It has to be capable of expressing or deriving some very general principles in relation to living systems. Since these operate on multiple scales, from the molecular to the whole organism and even beyond (organisms are not Turing machines; they are open systems), the ability to relate mathematical analysis at different scales is important. This is the reason why I think it may be important to investigate the application of the principles of scale relativity (Nottale, 2000; Auffray and Nottale, 2008). These were developed originally by the astrophysicist and relativity theorist Laurent Nottale (2000) as an extension of the relativity principles in general. There is an obvious correspondence here between such an approach and what I have called the theory of biological relativity (Noble, 2008). Scale relativity is a controversial development in physics and it is far too early to say whether the mathematical tools it develops can be applied to extend the equations we use in systems biology. It also makes some assumptions about space and time that people may find surprising, for example, that space-time is fractal. But remember $\sqrt{-1}$ (see Chapter 1). That also has no obvious 'existence' in what we consider to be 'normal' space and time. But, as a mathematical tool, it has turned out to be extremely fruitful. As quantum mechanics theory also recognises, the utility of a mathematical theory has little to do with whether we find it easy to conceptualise, or indeed whether it itself represents reality, whatever that is conceived to be.

Second, I suspect that the mathematics required for multiscale systems biology will need to explain attractors in the development of biological systems. The oscillators underlying the heart rhythm are a good example. The robustness of the system can be expressed by saying that, almost wherever the system may be at any time, it tends towards the oscillator as an attractor. The existence of the attractor is far more explanatory than any particular differential equation model since it is far more general, even if the equations for it are not capable of closed-form solutions.

Third, we need more general mathematical/computational theories of evolution. The 'theory of evolution' is not so much a theory as understood in the physical sciences. It is more a description, essentially a history, of what has happened. Biologists – except for a few fundamentalists – no longer argue about the fact of evolution. They argue fiercely though about the process and the mechanisms involved. We won't be able to resolve some of those arguments without some quantitative ways of representing what we think has happened. There are reasons for thinking that the process may be predictable since mutations are far from random (Stern and Orgogozo, 2009).

Fourth, we need to consider the arguments of those who are seeking non-algorithmic forms of mathematics to describe biological systems (Simeonov, 2010).

9.4. Conclusions

9.4.1. *Pluses*

The articles relevant for this chapter, and *The Music of Life* itself, represent a major shift in the way in which we think about biology. I am far from being the only person to think this way. Jablonka and Lamb (1995, 2005), Shapiro (2005, 2009b, 2009a), Margulis (1981), Keller (2000, 2002) and Dupre (1993) are just some of the other authors who are reinterpreting the fundamentals of biological science. And the most recent developments in evolutionary (Pigliucci and Müller, 2010) and genetic and developmental (Beurton *et al.*, 2008) theory also do so. I would like to think that *The Music of Life* is contributing not only to the shift within science itself but also in the battle for the hearts and minds of the educated public and the school students from whom the future biological scientists will be drawn. They all deserve better than the one-sided picture frequently presented to them.

9.4.2. *Minuses*

Nevertheless, we are a long way from achieving the goals. Systems biology still stands in need of a good definition. It could easily lead to the same problems that have befallen the Human Genome Project, with overblown claims and promises simply not being delivered. As my imagined Silmans contemplate the elephant, I would guess that we are at least a century away from anything as audacious as the Virtual Physiological Human. That is not to say that there won't be valuable insights along the way. It is to say that the project is enormous.

With regard to *The Music of Life*, the main minus that I see is that, despite its attempt to remove or replace one metaphor after another, some central metaphors remained untouched that should have been exposed for the problems they cause. The most blatant example is the word 'code', which is used throughout the book without a single warning. I would now replace nearly all uses of phrases like 'genes code for proteins' by 'genes form templates for proteins'. A template is not an instruction. It is simply used, just as an instrument maker uses templates to cut his wood or metal to the right shape. The word 'code' encourages people, wrongly, to think of the genome as a set of instructions. There are similar problems with the word 'information'. 'Every metaphor produces its own forms of prejudice' (*The Music of Life*, page 142).

9.4.3. *Contribution to systems biology*

When I first decided on the chapters under which this book would be organised, I tried to make them 10. There is something satisfying about 10 chapters, which is why *The Music of Life* is structured in that way. It also fitted naturally into the development of the successive implementations of the musical metaphor.

Before deciding to write the postscript to this book ('The Artist Disappears?'), I looked for something equivalent poetically to the ox herder parable. Could I end on such a note? In the end, I decided that would be a metaphor too far. The postscript can serve that purpose.

I close this chapter with the comment of Marc Kirschner (2005) when addressing an audience of young scientists about systems biology:

> You are probably running out of patience for some definition of systems biology. In any case, I do not think the explicit definition of systems biology should come from me but should await the words of the first great modern systems biologist. She or he is probably among us now.

The readers of this book may also like to think of themselves as possibly including the 'first great modern systems biologist'.

References

Auffray, C. and Nottale, L. (2008) 'Scale relativity theory and integrative systems biology 1. Founding principles and scale laws', *Progress in Biophysics and Molecular Biology*, 97, pp. 79–114.

Batchelor, S. (1994) *The Awakening of the West*. Berkeley: Parallax Press.

Batchelor, S. (1997) *Buddhism without Beliefs*. London: Bloomsbury Publishing.

Batchelor, S. (2010) *Confession of a Buddhist Atheist*. New York: Spiegel and Grau.

Beurton, P. J., Falk, R. and Rheinberger, H.-J. (ed.) (2008) *The Concept of the Gene in Development and Evolution: Historical and Epistemological Perspectives*. Cambridge: Cambridge University Press.

Brenner, S., Noble, D., Sejnowski, T., Fields, R. D., Laughlin, S., Berridge, M., Segel, L., Prank, K. and Dolmetsch, R. E. (2001) 'Understanding complex systems: Top-down, bottom-up or middle-out?', In *Novartis Foundation Symposium: Complexity in Biological Information Processing*, pp. 150–159. Chichester: John Wiley.

Crick, F. H. C. (1994) *The Astonishing Hypothesis: The Scientific Search for the Soul*. London: Simon and Schuster.

Dawkins, R. (1976, 2006) *The Selfish Gene*. Oxford: Oxford University Press.

Dawkins, R. (1982) *The Extended Phenotype*. London: Freeman.

Dawkins, R. (2003) *A Devil's Chaplain*. London: Weidenfeld and Nicolson.

Dupré, J. (1993) *The Disorder of Things*. Cambridge, Mass: Harvard.

Eccles, J. C. (1953) *The Neurophysiological Basis of the Mind. The Principles of Neurophysiology*. Oxford: Oxford University Press.

Egan, T., Noble, D., Noble, S. J., Powell, T., Spindler, A. J. and Twist, V. W. (1989) 'Sodium-Calcium exchange during the action potential in guinea-pig ventricular cells', *Journal of Physiology*, 411, pp. 639–661.

Gombrich, R. (2009) *What the Buddha Thought*. London: Equinox.

Hacker, P. M. S. (2011) 'The sad and sorry history of consciousness: Being among other things a challenge to the "consciousness studies community"', In *Human Nature* (ed. Sandis, C.). London: Royal Institute of Philosophy.

Hsu, E. (1999) *The Transmission of Chinese Medicine*. Cambridge: Cambridge University Press.

Jablonka, E. and Lamb, M. (1995) *Epigenetic Inheritance and Evolution. The Lamarckian Dimension*. Oxford: Oxford University Press.

Jablonka, E. and Lamb, M. (2005) *Evolution in Four Dimensions*. Boston: MIT Press.

Keller, E. F. (2000) *The Century of the Gene*. Cambridge, Mass: Harvard University Press.

Keller, E. F. (2002) *Making Sense of Life. Explaining Biological Development with Models, Metaphors and Machines*. Cambridge, Mass: Harvard.

Kim, S.-H. (2006) 'Blood and bone: A comparative study of body metaphors in Korean and British print media', In *English Language and Literature*, p. 319. Oxford: University of Oxford.

Kirschner, M. (2005) 'The meaning of systems biology', *Cell*, 121, pp. 503–504.

Kohl, P., Crampin, E., Quinn, T. A. and Noble, D. (2010) 'Systems biology: An approach', *Clinical Pharmacology and Therapeutics*, 88, pp. 25–33.

Lakoff, G. and Johnson, M. (1980) *Metaphors we Live by*. Chicago: University of Chicago Press.

Margulis, L. (1981) *Symbiosis in Cell Evolution*. London: W. H. Freeman Co.

McClintock, B. (1984) 'The significance of responses of the genome to challenge', *Science*, 226, pp. 792–801.

Noble, D. (1989) 'Intentional action and physiology', In *Goals, No Goals and Own Goals* (ed. Montefiore, A. C. R. G. and Noble, D.), pp. 81–100. London: Unwin-Hyman.

Noble, D. (1990) 'Biological explanation and intentional behaviour', In *Modelling the Mind* (ed. Said, K. A. M., Newton-Smith, W. H., Viale, R. and Wilkes, K.), pp. 97–112. Oxford: Oxford University Press.

Noble, D. (2004) 'Qualia and private languages', *Physiology News*, 55, pp. 32–33.

Noble, D. (2008) 'Claude Bernard, the first systems biologist, and the future of physiology', *Experimental Physiology*, 93, pp. 16–26.

Noble, D. (2008) 'Genes and causation', *Philosophical Transactions of the Royal Society A*, 366, pp. 3001–3015.

Noble, D. (2010) ' "Letter from Lamarck" ', *Physiology News*, 78, p. 31.

Noble, D. (2010) 'Biophysics and Systems Biology', *Philosophical Transactions of the Royal Society A*, 368, pp. 1125–1139.

Noble, D. (2011a) 'Differential and integral views of genetics in computational systems biology', *Journal of the Royal Society Interface Focus*, 1, pp. 17–15.

Noble, D. (2011b) 'Neo-Darwinism, the Modern Synthesis, and Selfish Genes: Are they of use in physiology?' *Journal of Physiology*, 589, pp. 1007–1015.

Noble, D. and Vincent, J.-D. (1997) *The Ethics of Life*. Paris: UNESCO.

Nottale, L. (2000) *La relativité dans tous ses états. Du mouvements aux changements d'échelle*. Paris: Hachette.

Pigliucci, M. and Müller, G. B. (ed.) (2010) *Evolution - The Extended Synthesis*. Cambridge, Mass: MIT Press.

Schrödinger, E. (1944) *What is Life?* Cambridge: Cambridge University Press.

Shapiro, J. A. (2005) 'A 21st century view of evolution: Genome system architecture, repetitive DNA, and natural genetic engineering', *Gene*, 345, pp. 91–100.

Shapiro, J. A. (2009a) 'Letting *E. coli* teach me about genome engineering', *Genetics*, 183, pp. 1205–1214.

Shapiro, J. A. (2009b) 'Revisiting the Central Dogma in the 21st Century', *Annals of the New York Academy of Sciences*, 1178, pp. 6–28.

Sherrington, C. S. (1940) *Man on his Nature*. Cambridge: Cambridge University Press.

Simeonov, P. L. (2010) 'Integral Biomathics: A post-Newtonian view into the logos of bios', *Progress in Biophysics and Molecular Biology*, 102, pp. 85–121.

Stern, D. and Orgogozo, V. (2009) 'Is Genetic Evolution Predictable?' *Science*, 323, pp. 746–751.

Wada, S. (2002) *The Oxherder*. New York: George Braziller.

Chapter 10

65 Years On: *Understanding Living Systems*

10.1. Opposing Metaphors in Biology

Publishing *The Music of Life* in 2006 was my first open venture into the, often fierce, controversies surrounding theories of evolutionary biology. That book established that opposing metaphors for genes in *The Selfish Gene* (Dawkins, 1976) do not lead to any empirical test. There is no biological experiment that could either validate or invalidate the metaphorical parts of the theory (Chapter 9, and pages 17–22 of *The Music of Life*). It therefore laid the foundations for my later work deconstructing that theory to show that it is not empirical science. I think the metaphorical parts are best represented as a tautology (Noble, 2013; Noble and Noble, 2021). At least some aspects of the theory are necessarily true regardless of which metaphorical description we use. Dawkins actually welcomes this description of his work:

> *The Co-operative Gene* would have been an equally appropriate title for this book, and the book itself would not have changed at all. I suspect that a whole lot of mistaken criticisms would have been avoided. (Dawkins, 2016, p. 347)

A key test of a tautology is precisely that completely opposing descriptions make no difference to the apparent logic. In this case, since it doesn't matter whether genes are selfish or cooperative, the logic of the theory will not tell us which is correct. That doesn't mean, of course, that there is no other component of the theory that might be true or false. As an example, he also writes the following:

> Another good title would have been *The Immortal Gene.* As well as being more poetic than "selfish", "immortal" captures a key part of the book's argument. The high fidelity of DNA copying — mutations are rare — is essential to evolution by natural selection. (Dawkins, 2016, p. 347)

This statement makes an empirical claim. But the problem here is that this fidelity does not derive from DNA alone! It is imposed by the living cell, which corrects many errors produced by self-replication. The living cell and the DNA are bound together in transmitting to the next generation. We could then just as well ascribe immortality to the cell. In fact, that might be more accurate. I suspect that cells evolved before DNA. Every germ cell in our bodies could then be said to be an 'immortal' line from the first living cells.

So, what are the relevant facts about replication? The ability for DNA on its own to 'replicate like a crystal' is accurate only to around 1 part in 4,000 to 10,000 nucleotides. Thus, inanimate systems based on nucleotides finding their own location within the replicating sequence (which is how crystals replicate) only achieve this degree of accuracy (Schulman, Yurke and Winfree, 2012; Deck, Jauker and Richert, 2011). Yet, our genomes are 3 billion base pairs long. The error rate of crystal-like replication would produce nearly 1 million errors. The replicator cannot therefore be separate from its vehicle. It is totally dependent on it. The difference, and the dependence on the living cell, is crucial since this is what allows living organisms to change their genes when they need to do so, as they did during the COVID-19 pandemic. It is the living organism that has the power to do that. If DNA could have been an automatic self-replicator, none of this crucial function in living organisms would have been possible. The ability of organisms to change genes when needed therefore runs deeply counter to the argument of *The Selfish Gene* since the separation between the replicator and its vehicle is a central foundation stone. Beyond the metaphors, therefore, there is, after all, an empirical test of a fundamental assumption of the theory of *The Selfish Gene*. DNA cannot be a sufficiently accurate *self*-replicator.

To fully understand this point, we also need to clarify the nature of causation in relation to DNA and why it cannot be an *active* cause, which I will do later in this chapter.

10.2. Early Encounters with Evolutionary Biologists

First, I need to explain when and why I first worked on evolutionary biology. Since publishing *The Music of Life*, I have published over 50 articles and two more books in this controversial field in just two decades.[1] Why was there such an apparently large and sudden transition? And seemingly far away from my original field of science, the heart?

The reality is very different from the appearance. My interest in and involvement in evolutionary biology is a very long-standing one. I was introduced to *The Modern Synthesis* view of evolution during my school biology education in London, based on *Animal Biology* (Grove and Newell, 1945), the first edition of which was published in the same year as Julian Huxley's (1942) *Evolution: The Modern Synthesis*, and was faithful to that synthesis. As a medical student at University College London in the 1950s, I attended the lectures by the famous Zoologist JZ Young, author of *The Life of Mammals* and other important books that influenced me to abandon a career in medicine and become a physiological scientist.

But the crucial trigger was one of my first additional assignments by Oxford University after I moved there in 1963. I was appointed examiner in 1966 of the doctoral thesis of a research student, who has since become a highly successful author of worldwide fame, Richard Dawkins.[2] The thesis was not on the work that led to *The Selfish Gene*. That book came out 10 years later in 1976. The 1966 thesis was based on research on bird behaviour, supervised by the Nobel Prize-winning ethologist Nikolaas Tinbergen. That research used mathematical analysis to understand time series based on observations of the timing and duration of different

---

[1] www.denisnoble.com/evolution-publications.

[2] 1966 was also the year in which I was first connected with at least the name 'Darwin': I was invited to give the Darwin Lecture to the British Association for the Advancement of Science. I did not know then that 40 years later I would be defending Darwin's theory of evolution against the neo-Darwinists!

activities of birds, such as foraging, feeding, preening, sleeping and mating. I was asked to be an examiner, to join an ethologist from another university as the co-examiner, because I was one of the very few biological scientists in Oxford at that time known for using mathematics in biology in a more than trivial way. This itself was already a connection with theories of evolution since Tinbergen was also the formulator of what became known as Tinbergen's four questions.[3]

The author of *Evolution: The Modern Synthesis*, Julian Huxley (1942), and the evolutionary biologist, Ernst Mayer (1982), had already introduced 3 of the 4 questions: behavioural adaptive functions, phylogenetic history and physiological mechanism. Tinbergen added the fourth: ontogenetic/developmental history (Tinbergen, 1963). For a complete biological analysis of a species, all four questions need to be answered. Dawkins's work was clearly in the first category. The timelines of behavioural elements were being analysed using series transformations, including the Laplace transforms.

As it happens, at that time I knew nothing about Laplace transforms! But I rapidly taught myself before the viva by buying books on them. That helped me immensely when writing *Electric Current Flow in Excitable Cells* (see Chapter 3). There was nothing about selfish genes in the thesis, but 1966 happens to also be the year when George Williams (1966) wrote his book *Adaptation and Natural Selection*, which was later, by Dawkins's own admission, to become the basis of *The Selfish Gene* (Dawkins, 2019).

By the time of the publication of *The Selfish Gene*, I had become the head (Praefectus) of Balliol's Graduate Centre at Holywell Manor. I immediately invited Richard Dawkins to talk about his book at one of our evening seminars. I also invited two philosophers, Anthony Kenny, author of *Action, Emotion and Will* (Kenny, 1963), and Charles Taylor, author of *The Explanation of Behaviour* (Taylor, 1964), to join the evening to make it into a debate.

The large Praefectus' study at Holywell Manor was packed out for the event. My strategy was to initiate a debate between the two scientists, Dawkins and me, and the two philosophers, Kenny and Taylor. But that was not how it turned out. Early in the evening, Anthony Kenny raised the question, 'If all I knew of the English language was its alphabet, surely I would not be able to understand the works of Shakespeare?' I shall never

---

[3] https://en.wikipedia.org/wiki/Tinbergen%27s_four_questions.

forget Dawkins's reply: 'Well, I am not a philosopher. I am a scientist. I am only interested in Truth'.

Kenny's point was, of course, to raise the question about how the DNA sequences come to have meaning in order to enable physiological functions, which is also why the question was not a purely philosophical one. Functions can be proven or disproven empirically. Yet, meaning and function surely cannot come from a sequence alone, since it needs an interpreter, such as a living cell. Dawkins did not address Kenny's point. 'I am not a philosopher' is frequently used by Dawkins, and it is an attitude I encountered a lot in my undergraduate studies since it is used by many reductionist scientists. The biologists and chemists who taught me at University College London in the 1950s would often make fun of philosophy, portraying it as a discipline that has nothing to do with science and never makes progress.[4] The co-discoverer of the double helix, James Watson, often repeated his mantra, 'There are only molecules, everything else is sociology'. As for philosophy, he is quoted as saying, 'I do not like to suffer at all from what I call the German disease, an interest in philosophy'. Jerry Coyne, author of *Why Evolution is True* (Coyne, 2009), frequently treats philosophy, and particularly the philosophy of mind, with contempt (Coyne, 2014).

I had become sceptical of these general attacks by reductionist scientists on the discipline of philosophy through attending the graduate classes of Stuart Hampshire, Professor of Philosophy at UCL, where I learnt just how little I knew of the ease with which unthinking people can fall fowl

---

[4]One of those was Lewis Wolpert who later organised a Novartis Foundation Symposium on 'The limits of Reductionism' (Wolpert, 1998). His opening statement at the meeting was 'there are no limits!' A more complete statement is at the end of the published book (Bock and Goode, 1998). J Gray had asked Lewis, 'Lewis, tell us why you care so deeply that "reductionism" should always win!' Lewis Wolpert replied, 'Because I think that there is no good science that doesn't have a major element of reductionism in it, and holism is dead. Reductionism has been amazingly successful. The limits of reductionism are only a reflection of our ignorance. The world, however, is a complicated place, and if you think something cannot be understood, my advice is "be patient", because a reductionist explanation will eventually emerge. It may however take a long time with systems as complex as the brain and may require quite new concepts and technologies. We may even have to change or at least confront what we mean by understanding in biology'. I disagree fundamentally with Wolpert's statement here. Neither I nor anyone else can even solve the differential equations for molecular-level processes without the boundary conditions set by higher levels of organisation in living systems.

of misunderstanding language, our basic means of communication to each other, and without which science could never even be communicated. I was asked to present a paper to the class. At the end, Stuart Hampshire recommended that I read Spinoza. That eventually led to me discovering the important letter in Latin that Spinoza sent to the Royal Society, which led me to his statement of the principle of biological relativity. The details of that discovery can be found in *Dance to the Tune of Life* (Noble, 2016, pp. 165–168).

When I examined Dawkins's thesis, I was also engaged in discussions with the philosopher Charles Taylor on his book *The Explanation of Behaviour*, leading to a debate with him (Noble, 1967a, 1967b; Taylor, 1967), eventually published in the philosophical journal, *Analysis*, just a year (1967) after examining Dawkins. I have already highlighted the significance of those interactions with Taylor in Chapter 8, but it is important also to do so in this concluding chapter since they relate to the problems of causation in biology. That was why I knew Taylor's work could be relevant to discussing *The Selfish Gene* in 1976.

But Dawkins made it clear that he was not interested at all in addressing Kenny's or any other philosopher's argument. Yet, I was already well primed to see why it was *necessary* for scientists to address the issue of meaning since, in the opening (Noble, 1967a) of my published debate with Charles Taylor, I had used a similar approach to that of Dawkins by proposing that micro-level, such as molecular level, explanations could be complete. That would necessarily exclude any higher-level explanation from adding anything of value. That would require there to be meaning. Both Tony Kenny and I initially thought this was a knockdown argument. So, we were intrigued, to say the least, by Taylor's reply (Taylor, 1967). He granted the case, but nevertheless replied with a very insightful comeback. He agreed that might be true in any particular case, but what would be true if one considered *a whole series of cases*? Might it not then be true that the higher level shows order while the lower level does not? In that case, how could the lower level explain that order, since the order clearly did not exist at that level?

Much later, I found that there are many examples of this situation in nature. Different species of high-altitude birds all have the characteristic that their oxygen–haemoglobin dissociation curves are functionally shifted to bind oxygen more strongly so that they can capture oxygen efficiently even when the oxygen partial pressure is low. But the precise genetic variations by which this is achieved seem to be randomly different between the species (Natarajan *et al.*, 2016). This is a version of a fact we will encounter later in this chapter: there can be many different genetic patterns that can achieve the same higher-level function, just as there are several different processes by which pacemaker activity can be maintained in the heart. There lies the strong connection between my earlier work on the heart and what I am doing now on evolutionary biology. Without that firm experimental background in robustness and redundancy in biology, I doubt whether I would have made the move into the field of evolutionary biology when I published *The Music of Life*.

10.3. Ignoring Philosophy is a Risky Strategy

What these philosophical arguments and experimental biological examples show is that it is a risky strategy to completely rule out the use of philosophy in science. There are philosophical/linguistic traps whenever we try to put raw scientific data into an interpretive, explanatory context, since we must use language, where the interpretive understanding of metaphors (*selfish* genes?), irony (Central *Dogma*?) and many other non-literal forms of language becomes essential. Some modern genomics researchers even try to avoid theory and explanation completely. They argue that the only function of science is to accumulate data (Yanai and Lercher, 2020). Those authors even regard a hypothesis as a liability. That approach has led to deep misunderstandings of causation in biology (Felin *et al.*, 2021). Simply accumulating data must at some point be subject to experimental or theoretical test if it is to become science rather than just the scientific equivalent of stamp collecting. Anyway, there is, inevitably, a theory behind association studies. This is the hypothesis that association studies will be useful in the prediction of disease and in discovering its cures. Why else would government and other funding agencies have spent huge sums of money on this work?

10.4. How to Test the Predictive Power of Genomics Data Experimentally

What would be such a test in the case of genomic data? A team at University College London showed precisely how to do that. Just ask the question, 'From that data can we usefully predict what diseases someone might suffer from in the future?' So, they subjected the scores in the polygenic scores repository (https://www.pgscatalog.org/) to the test of whether they can be relied on to predict major fatal diseases, such as cancer or cardiovascular disease. They employed the same statistical criteria used to decide whether a new drug may be approved for public use by the Food and Drug Administration (FDA): do the data available on what the drug does favour its therapeutic effect without incurring too many negative effects? The results of that test were published in 2023 in the journal *BMJMedicine*:

> Polygenic risk scores performed poorly in population screening, individual risk prediction, and population risk stratification. (Hingorani *et al.*, 2023, p. 1)

This outcome is shocking since the practical reason justifying the enormous cost of mass genome sequencing of large populations was precisely that, within 10 years of doing so, cures for the major fatal diseases of mankind would be found (Collins, 1999). That, clearly, has not happened even 25 years later. It is important to ask why? And to ask what would be a more successful alternative? It is urgent to do so since the major fatal diseases of old age are now dominating expenditure on health and social care budgets. Misunderstanding living systems in the way that purely reductionist explanations do is not only bad theory but it also leads to extremely costly strategic mistakes. In retrospect, it is now clear that it was the wrong strategy to throw nearly all the financial medical research resources into genomics. It would have been better for humanity if we had also continued the successful physiological research that had led to effective development of new medications in earlier decades. The research of my own team led directly to the development of an effective treatment for dangerously rapid heart rhythm, the Servier company's drug ivabradine (DiFrancesco and Camm, 2004). That development required a combination of molecular-level understanding of the genes and proteins involved and an integrative demonstration of why that could slow cardiac rhythm without the risk of stopping it completely. Both integrative and reductionist research had to be used to achieve this outcome. Yet, many genomics researchers seem to think that identifying genes causing a problem and then repairing or replacing them will be sufficient. In polygenic functions, it is clearly unlikely to work without taking the integrative characteristics into account.

10.5. Robustness in Biology is Widespread

After my cardiac research team had identified the multifactorial backup processes protecting heart rhythm from sudden arrest, I looked for other examples of such robustness in living systems. The answer is that, with the exception of the well-known monogenetic diseases, such as the skeletal muscular disease, cystic fibrosis and other monogenetic diseases affecting around 4% of the human population (Bavisetty, Grody and Yazdani, 2013; Rodwell and Ayme, 2014), robustness is a widespread feature of the important biological functions that keep us alive. It means that we must now investigate those integrative physiological processes at higher levels of biological organisation if we are to succeed in finding the cures that genome sequencing has failed to do.

The problem here is that integrative physiological research was sidelined and underfunded for more than 40 years. In some advanced countries, including the UK, many departments in universities devoted to such research were closed. Those skills were lost for at least a generation. In effect, the world was persuaded to bet its future health care on the seductive hope of quick genetic solutions (within 10 years!) at the expense of seeking real understanding of how living systems work. Counting genes and their association scores is the equivalent of identifying the pixels in an image of a message instead of understanding the message itself. That is another way of expressing the question that Kenny put to Dawkins. The full import of that question will now be clear. It was not a purely philosophical question.

It will now take decades of education and the encouragement of a new generation of interdisciplinary researchers if we are to recover the momentum that such research had more than a generation ago. Cures for the complex multifactorial omnigenic diseases that the world now needs for its ageing populations will not be a 10-year sprint. Given the combinatorial explosion involved in higher-level interactions in living systems, I believe we are looking at least a couple of generations to recover from our present gene-centric impasse. The search for cures will need to focus on restoring function, not just on restoring templates for proteins.

10.6. Evolution of Plants: A 100-Million-Year Experiment in Multi-Component Medicine?

Evolutionary biology will have a major contribution to make to such research. That is because developing multifactorial nutritional treats that can keep organisms healthy, and even cure complex health problems, has been the outcome of an evolutionary experiment that has now lasted at least 100 million years since the evolution of flowering and fruiting plants, with some estimates suggesting even longer (Barba-Montoya *et al.*, 2018), at least 200 million years. Over that long period of time, plants have depended on mobile organisms, animals, to spread their seeds and eventually also to widely cultivate their (the plant's) development and evolution. To achieve this outcome, plants have evolved chemical combinations in their edible parts that are attractive and, *in those combinations*, more effective than pure single chemicals can ever be. That is partly why most plant products can be cultivated for nutrition.

That evolution over millions of years is an extraordinary experiment that evolution has carried out. There is no other way humanity could itself have arrived at those combinations. We humans have not existed for more than a tiny fraction of those 100 million years, and our lifetimes are far too short. We can now, however, learn from what plants have developed as our multifactorial nutrition. We could study the synergies and incompatibilities between the chemicals in plant products. In a somewhat haphazard way, that is what herbalists have been doing for thousands of years.

Remember, too, that large sections of the world's pharmacopoeias originally developed from botanical sources. But we have largely done that by trying to identify *single* chemicals (the classic Western drug) that will be curative, when what we need is *functional combinations*. Plants, in their multi-million-year evolution, have already arrived at those solutions without, of course, understanding why they work. It is now up to us to perform the research needed to acquire that understanding. My research team's work, using mathematical modelling of complex physiological interactions, combined with molecular level investigations, has already shown the way in which that can be achieved.[5]

10.7. Discussion with Richard Dawkins, 2022

Multifactorial robustness is not only important for why living organisms are successful in maintaining themselves but also because it helps us understand causation in living systems. That must also be a major

---

[5] See www.denisnoble.com/systemsproject.

objective of a new approach to biology. It cannot be emphasised too much that causation is not the same as association.

The idea that it might be the same was the central issue in a high-profile discussion between me and Richard Dawkins in 2022. It was held at the annual Festival of the Institute of Art and Ideas held in Hay-on-Wye. The viewing figures of the various internet versions of that recording now total well over 1 million. That reflects the huge level of public interest in Dawkins's books. Justly so, since I see him as the greatest and most successful popular exponent of the neo-Darwinism theory of evolution. That theory is based on Charles Darwin's first theory of Natural Selection, published in *The Origin of Species* in 1859. Natural Selection was subsequently integrated with Mendelian genetics during the first half of the 20th century to produce the Modern Synthesis (Huxley, 1942).

But neo-Darwinism is not a correct representation of Darwin's later work, published in 1868, *The Variation of Animals and Plants under Domestication*, and 1871, *The Descent of Man* and *Selection in Relation to Sex*. In those books, Darwin laid out the reasons due to which he was convinced that at least three other processes needed to be added to natural selection: sexual selection, pangenesis and the inheritance of use and dis-use processes in the body. He worked with a young physiologist, George Romanes, on these ideas during the last 10 years before he died in 1882. Romanes published the outcome of those interactions in his 3-volume work, *Darwin and After-Darwin* (Romanes, 1886). That work included an important theory of Physiological Selection in the third volume (Noble and Phillips, 2024).

These differences between neo-Darwinism and Darwin's more fully developed later ideas are important because recent work has shown that Darwin's additional processes can operate in much the same way as Darwin proposed. I do not think therefore that Darwin would ever have accepted the neo-Darwinists' presumption and implication that he would have agreed with them.[6] He disagreed *strongly* with the scientists I see as the founders of neo-Darwinism in the 19th century: Alfred Russel Wallace, Francis Galton and August Weismann (Desmond and Moore, 1991; Noble and Noble, 2025).

---

[6]For that reason, I believe it is not strictly correct to use Darwin's name in 'neo-Darwinism'. It creates deep confusion. I frequently get misrepresented as either being opposed to Darwin or that I must be a neo-Darwinist since I often defend Darwin.

10.8. Causation in Biology

We are now ready to return to the problem of causation in biology and why I directly disagreed with Richard Dawkins in our discussion on gene causation at the IAI Festival in 2022. That is important because he has repeated his position on gene causation in his recent book, *The Genetic Book of the Dead*. In Chapter 8 of that book, he replied to a citation from my 2016 book, *Dance to the Tune of Life*:

> This book will show you that there are no genes "for" anything. Living organisms have functions which use genes to make the molecules they need. They are not active causes. (Dawkins, 2024, pp. 176–177)

His reply in *The Genetic Book of the Dead* is as follows:

> Successful genes are those with a statistical tendency to inhabit bodies that are good at surviving and reproducing. And they enjoy that statistical tendency, positive or negative, by virtue of the *causal* influence they exert over bodies. So we have arrived at the reason why it was profoundly wrong to say that genes are not active causes. Active causes is precisely and indispensably what they must be. If they were not, there would be no natural selection and no adaptive evolution. (Dawkins, 2024, p. 183)

Note the use of '*active* causes' (my emphasis), since readers may be surprised by the fact that I would not have disagreed with this part of Dawkins's argument if he had not used the word 'active' before 'causes'. That is why in *Dance to the Tune of Life* I carefully distinguished between the different forms of causation (see Chapter 6 in that book). Specifically, I distinguished active causes, frequently represented mathematically by differential equations, from passive causes arising from structure and form. A confusion between 'active causes' and the implied opposite 'passive causes' is central to my disagreement with Dawkins. Note also that he does not refer to this particular distinction. Yet, it is crucial to understanding the point made in the citation from *Dance to the Tune of Life*. Without understanding the difference between active and passive causes, the full import of my argument that association is not the same as causation cannot be understood. Moreover, to understand that distinction, the subtitle of my book is important. The distinction is also crucial to understanding why genes cannot form the complete blueprint for life (Noble, 2024).

The subtitle of my book is *Biological Relativity*. Dawkins's comment on that is also relevant to sorting out the muddle here:

> *Dance to the Tune of Life* has the subtitle 'Biological Relativity'. Noble's usage of 'relativity' has only a tenuous and contrived connection with Einstein's. (Dawkins 2024, p. 177)

It might conceivably have 'only a tenuous and contrived connection' with Einstein's theory of *Special* Relativity, but that is not the form I derived the idea from. I believe though that the principle of Biological Relativity has a deep connection with the relativity of causation in Einstein's theory of *General* Relativity, which is an important form of relativity in relation to causation in the universe. The full reasons for that distinction are outlined in Chapters 1 and 6 of *Dance to the Tune of Life*, but I will briefly explain those reasons here before I address the distinction between active and passive causes in the argument about how genes have a causal effect.

10.9. Origins of Biological Relativity in Relation to the General Principle of Relativity

The theory of General Relativity can be seen as deriving from an even more general principle, the *general principle of relativity*[7], in which the equations of physics have the same form in all admissible frames of reference. This can be interpreted as a relativity of causation. In its application in Einstein's theory of General Relativity, it is expressed in the two-way causation between space-time and the structures within space-time. Those structures distort space-time so that light travels in curved space, thus giving rise to effects like gravitational lensing. In turn, curved space-time influences how objects and electromagnetic radiation can move within it.

This means that structural form in the universe *is itself a form of cause*. This idea was enunciated by Aristotle over 2,000 years ago. He understood that it was important to distinguish between the way in which form *passively* influences what happens just by existing and active

---

[7]en.wikipedia.org/wiki/principle_of_relativity.

causation, which he called *efficient* causation. Both of these forms of causation are required when we solve the differential equations for *efficient* (*active*) causation since those equations have no specific solutions unless we specify the initial and boundary conditions. Those boundary conditions are necessarily derived from the form of the structures within which active causation happens. The relativity of causation in biology is therefore a necessary mathematical truth, without which no formally closed solutions are possible.

10.10. What Kind of Cause Can be Attributed to DNA?

Now, we can answer this question: what form of causation may DNA be involved in? We know the answer to that question since DNA on its own does nothing *actively*. Chemically, it is a *passive* molecule. In order to exert any influence on organisms, its sequence needs to be read. That is when its form, the sequence of nucleotides, becomes transferred to a different form of nucleotide sequence, RNA. RNAs differ from DNA in that they can also be active causes, for example, as enzymes. This is one of the reasons why many evolutionary biologists believe that an RNA world may have preceded the evolution of DNA. RNA viruses may then be a remnant of 'living' organisms in that form.

The RNA sequences are then used by ribosomes in living organisms to use the sequence pattern derived from DNA to generate the many different forms of polypeptides we call proteins. What is happening in this process is a mixture of passive causation by form with active causation by mechanism, i.e. precisely the combination required by the general principle of relativity. Without that combination, there could be no specific solutions to differential equations we may have derived from the active processes in living systems.

DNA does not therefore perform any of the active chemical processes involved. It needs the relevant part of its thread to be activated in order for transcription to RNA to be initiated, and it requires RNA polymerase to read the DNA, which forms a passive template for the creation of the RNA. When a cell is dividing, DNA requires the active participation of the living cell, which orchestrates a set of DNA cut-and-paste enzymes to correct a large number of errors as the copying process proceeds. The replicator (DNA) is not therefore independent of its vehicle (the living cell).

10.11. Relation Between Passive and Active Causes in Relativity

Biological Relativity is a formulation of the fact that passive causes provide conditions that enable the active causes (expressed as differential equations) to have solutions, just as in Einstein's theory of General Relativity the formal structure of space-time necessarily influences the movements of physical objects and therefore constrains the relevant differential equations for those movements.

This way of viewing the processes involved is also easy to visualise from the Hodgkin Cycle discussed in Chapter 1, Fig. 1.10. Downward causation from the electric field is a passive cause. It forms a boundary condition within which the dynamic molecular processes must proceed. It is not surprising therefore that the Hodgkin Cycle is one of the origins of the principle of Biological Relativity (Noble, 2022). Moreover, a Hodgkin cycle differs fundamentally from a purely sequential biochemical cycle, such as the Krebs (citric acid) cycle. In the case of a Hodgkin Cycle, the boundary and dynamic equations must be integrated *simultaneously* (Noble, 2022). In the case of Metabolic Cycles, the differential equations might be solved sequentially and this is central to understanding multi-level living systems.[8]

I can find nowhere in Dawkins's writing where this crucial distinction between passive and active causes is even acknowledged. Yet, without it, a complete understanding of causation in multilevel organisms is impossible. Anyone who reads *Dance to the Tune of Life* further than its introduction can hardly fail to see that it distinguishes between active and passive causes and that this distinction is central to the principle of biological relativity.

This insight is yet another area in which philosophical speculation was valuable. As Chapter 1 of *Dance to the Tune of Life* explains, there were many philosophical predecessors of relativity theories.

---

[8]It is of course possible to combine Hodgkin and metabolic cycles. The metabolic processes in mitochondria cease when the mitochondrion is depolarised, as also happens in bacteria, from which the mitochondria evolved. To represent this fact would require a simultaneous solution to equations for the Hodgkin cycle, up and down. A complete modelling of a mitochondrion would require this.

Most importantly, in their application to biology is the work of Spinoza. In a letter to the Royal Society in 1665, he writes the following:

> Let us imagine, with your permission, a little worm, living in the blood, able to distinguish by sight the particles of blood, lymph etc, and to reflect on the manner in which each particle, on meeting with another particle, either is repulsed, or communicates a portion of its own motion. This little worm would live in the blood, in the same way as we live in a part of the universe, and would consider each particle of blood, not as a part, but as a whole. He would be unable to determine, how all the parts are modified by the general nature of blood, and are compelled by it to adapt themselves, so as to stand in a fixed relation to one another.

Those particles of blood are constrained by the *form* of the vascular system. The principle is applicable to all organised systems in biology. Spinoza was, of course, the main opponent of Descartes's reductionist preformationism (Noble, 2016, pp. 165–168).

10.12. Creating Clones from DNA

A second important interaction in the 2022 discussion between me and Dawkins is the question of whether my DNA could be used to recreate me, or at least an accurate replication, a clone, of me.

It is fascinating to spell out what Dawkins may be proposing when he says that, in 10,000 years' time, a clone of me could be made from the information stored in my genome. He also admits that it would need an egg cell, but one from 'any woman' would do. So, how might this be achieved?

The relevant part of the 2022 transcript reads thus:

> Suppose somebody put Denis's genome in a Petri Dish. And keep it going for 10,000 years. Well, it wouldn't keep going. It would decay, as you rightly say. However, the information, it could be preserved on paper. You could actually write it down in a book, you could carve the ATC and G codons in granite and keep it for 10,000 years. And then in 10,000 years, type it into a sequencing machine, which we already have, and it would recreate an identical twin of Denis Noble. (Transcript: https://www.denisnoble.com/wp-content/uploads/2023/02/Transcript References.pdf, p. 4)

And the relevant part of Dawkins (2024) reads thus:

[in 10,000 years' time] they'll have the embryological knowledge to create a clone of whoever donated the genome in the first place (just a version of the way Dolly the sheep was made). Of course, the DNA information would need the biochemical infrastructure of an egg cell in a womb, but that could be provided by any willing woman. (Dawkins, 2024, pp. 179–180)

This assumes that it doesn't matter who provides the egg cell. I think it must.

My reply, therefore, was 'Where would you get my mother's egg cell as it was in February 1936 [when I was conceived]?'

All egg cells carry specific epigenetic marking from the mother and their environment and cytoplasmic properties that can influence the development of the embryo, while information is also transferred from the womb and the mother's body (McLaren and Michie, 1958; Bateson, 2001; Gluckman and Hanson, 2005, 2006). Cytoplasmic inheritance matters. That was demonstrated in a similar experiment performed on two species of fish at the Wuhan Fish Institute in China. A nucleus from a carp was inserted into the enucleated egg cell of a goldfish (Sun *et al.*, 2005). The fish that developed from this clone had a body shape and vertebral column number roughly halfway between that of a carp (about 36 vertebrae) and a goldfish (around 25 vertebrae). We cannot therefore assume that the development of the clone will be determined entirely by the nuclear genome of the transplanted nucleus. Furthermore, whether those epigenetic effects persist for more than a generation or two will depend on whether the environment is the same or different from that of the nucleus donor. Temporary environmentally induced epigenetic effects will die out. But if that environment persists, there is every reason to expect that the epigenetic changes will also persist. The environment in which an embryo develops is also important.

This is a key feature of the concept of Physiological Selection, as proposed by Romanes after Darwin's death (Romanes, 1886). Of course, neither Romanes nor Darwin knew about epigenetics. Romanes's speculation was that physiological processes (which we now know would depend on epigenetic changes) might prevent interbreeding between developing varieties. If so, then branching speciation could occur (Noble and Phillips, 2024). Physiological Selection (which now would inevitably include epigenetic changes) was the direction in which the decade-long collaboration between Darwin and Romanes was heading. But all they had to support the idea was Darwin's theory of pangenesis. The 'gemmules' he postulated to fulfil this role are what we now call extracellular vesicles or exosomes (Noble, 2019), which carry out the process of transmission to the germ line that Darwin postulated in 1868 (Phillips and Noble, 2024). That part of their speculation was therefore correct.

In this concluding chapter, I have not only updated the version of this book published in 2012 but I have also shown how the robustness of the pacemakers of the heart motivated my return to the field of evolutionary biology. The question that remains is, why was that return delayed for so long? What happened in 2004 when I began writing *The Music of Life*?

That was the year I retired from my chair in Cardiovascular Physiology. I no longer needed to apply for research grants to support the salaries and equipment of a cardiac research team. The many critical reactions to my work on evolutionary biology during the subsequent 2 decades reveal that caution was not only wise but also absolutely necessary! I wrote in my Introduction, 'Controversy: that is what changed my mind'. Controversies in my work on cardiac physiology pale in significance compared to the controversies over my work on evolution. That those controversies would have seriously impacted the funding for my research team can now hardly be doubted. I had witnessed the sad demise of many other physiological research teams in the UK, succumbing to this steamroller flattening of the very discipline necessarily needed now to recover from the gene-centric impasse.

10.13. *Understanding Living Systems*

Understanding Living Systems (Noble and Noble, 2023) is a book written with my brother, Raymond, as the lead author. Ray and I have collaborated for the last 8 years on many publications on evolutionary biology. That collaboration began immediately following a meeting in 2016 on *New Trends in Evolutionary Biology* held at the Royal Society, the papers for which were published in *Interface Focus* in 2017. My own contribution to that publication outlined what we call the 'Harnessing of Stochasticity' (Noble, 2017). We followed that concept up with a series of articles (Angelis *et al.*, 2019; Noble and Noble, 2017, 2018, 2019a,

2019b, 2020, 2021a, 2021b; Noble *et al.,* 2019), now collated together on my website, www.denisnoble.com as the 'Harnessing Collection'.[9]

Ray and I are both physiologists. He worked on neuroscience in Edinburgh, then on neonatal physiology at UCL (where he co-founded the Institute of Women's Health). I am a cardiac physiologist, having also worked first at UCL and now at Oxford. A question that has exercised both of us throughout our careers as physiologists is, 'What is the physiological basis of intentional agency in living organisms?' Orthodox evolutionary biologists tend to dismiss this question as precisely the kind of philosophy they denigrate, even to the extent of claiming not to 'be philosophers'. In Futuyma and Kirkpatrick's mammoth textbook, *Evolution* (Futuyma and Kirkpatrick, 2018), there is only one page (p. 20), out of more than 600, devoted to philosophy, which they dismiss with the concluding statement, 'The concept of purpose plays no part in scientific explanation'.

This sentence is simply a declaration. There is no attempt at explaining why this should be so. It is the equivalent of Dawkins's mantra, 'I am not a philosopher'. The authors rely on no one questioning 'why not?'

Ray and I fundamentally disagree with this dismissal of the relevance of philosophy. All living organisms anticipate their environment and what other organisms may do in that environment. As the Nobel Prize-winning immunologist, Gerry Edelman (1978; Edelman, Gally and Baars, 2011), realised many years ago, this is a physiological process, just as the antici-patory search for new DNA sequences that can generate a successful immunoglobulin is a physiological process. It involves a choice between the many sequences generated randomly by the immune system cells. There is clearly an anticipated target: the need to neutralise a new virus or bacterium. Similarly, organisms need to choose between their behavioural options when chasing prey or fleeing predators when they themselves are prey. Edelman called this Neural Darwinism, precisely because it involves a selection between alternative neuronal processes that may produce the required behaviour.

It is hard to understand why this should not be regarded as science. It does not involve a spooky 'ghost in the machine', nor does it require backward time causation (a goal in the future determining the behaviour now?). The anticipatory state of the nervous system *precedes or accompanies* the purpo-sive action. The anticipatory state is not later than the required behaviour.

---

[9]https://www.denisnoble.com/wp-content/uploads/2021/01/Harnessing-Collection.pdf.

In our 2020 article in the *Journal for General Philosophy of Science*, we explained 'How Might Intentional Agency Work Physiologically?' which was also the subtitle of the paper. The article was published in a philosophical journal because this question is precisely where philosophy and physiology meet. In a century in which creators of AI are busy building AI systems that can mimic this form of behaviour, it is ostrich-like (head in the sand) for scientists to pretend that anticipatory behaviour (which is what we mean by purposive behaviour) cannot be studied scientifically.

When we began writing *Understanding Living Systems*, we decided to meet this problem head-on. Our Preface ends thus: 'Understanding Living Systems involves understanding their agency'. Charles Darwin agreed with this viewpoint. In his 1871 book on *Selection in Relation to Sex*, he wrote as follows:

> When we behold two males fighting for the possession of the female, or several male birds displaying their gorgeous plumage, and performing strange antics before an assembled body of females, we cannot doubt that, though led by instinct, they know what they are about, and consciously exert their mental and bodily powers. (Darwin, 1871, p. 245 in Penguin Reprint, 2004)

He repeated this for the case of females:

> The exertion of some choice on the part of the female seems a law almost as general as the eagerness of the male. (Darwin, 1871, p. 257 in Penguin Classics reprint. 'Our emphasis.')

Moreover, we have entered an age when the ability of organisms to change the world in which they live is now threatening the survival of the very same species that is responsible for the change! That is why our book ends with a challenge to the young generations:

> It will require creative ingenuity to shift the culture of biology away from the misunderstandings of the twentieth century.... It will be for a new generation to discover and create their own culture fit for the challenges of the twenty-first century.
>
> They will have plenty of looming signposts to warn them what went wrong. Theirs will be a generation that must take responsibility for the way in which the earth's ecosystems need rescuing, even for our own species to survive.

Theirs will be the generation that faces the challenge of ageing societies, requiring medical science to find solutions to diseases of old age that do not readily yield to reductionist gene-centric solutions since those diseases are multifactorial. Only an integrative approach that understands those multifactorial interactions can possibly hope to address those diseases.

Theirs will be a generation that can try to recover from the damage to society that results from reductionist models of physiology and evolution that have metaphorically shaped ideas and models in fields as diverse as economics, sociology, philosophy, ethics, politics ... the list goes on because no aspect of today's society can have escaped dogmas like 'we are born selfish', 'they [genes] created us body and mind', 'it's in their DNA', and the myriad of other tropes of related types that we now use almost without thinking.

Those future generations will also need to rewrite the textbooks, not only because they see the virtue of 'let us therefore teach our children', but also because their politicians, economists, sociologists and philosophers will also need to find new strategies, in collaboration with biologists who can lead them out of the gene-centric impasse.

It is arguably a challenge the scale of which human society has never faced before.

WE WISH THEM ALL WELL

As I write the concluding lines of this book, I am also working on the modelling methods that may lead the way back into understanding how to discover treatments that respond to the challenge of multifactorial diseases (Noble and Hunter, 2020). Understanding living systems from an integrated multifactorial viewpoint is urgent, and just as vital as understanding the process by which organisms like us can be kept alive through the automatic process of their heartbeat rhythms.

Books Most Relevant to This Chapter

The Selfish Gene, Richard Dawkins (1976);
The Genetic Book of the Dead, Richard Dawkins (2024);
Dance to the Tune of Life, Denis Noble (2016);
Understanding Living Systems, Noble Raymond and Noble Denis (2023);
The Beating Heart, Robin Choudhury (2024).

References

Angelis, A. de, Hossaini, A., Noble, D., Noble, D., Soto, A., Sonnenchein, C. and Payne, K. (2019) 'Forum: Artificial Intelligence, Artificial Agency and Artificial Life', *RUSI Journal*, 164, pp. 120–144.

Barba-Montoya, J., Reis, M. d., Schneider, H., Donoghue, P. C. J. and Yang, Z. (2018) 'Constraining uncertainty in the timescale of angiosperm evolution and the veracity of a Cretaceous Terrestrial Revolution', *New Phytologist*, 218, pp. 819–834. doi:10.1111/nph.15011.

Bateson, P. (2001) 'Fetal experience and good adult design', *International Journal of Epidemiology*, 30, pp. 928–934.

Bavisetty, S., Grody, W. W. and Yazdani, S. (2013) 'Emergence of pediatric rare diseases: Review of present policies and opportunities for improvement', *Rare Diseases*, 1, Article e23579. doi:10.4161/rdis.23579.

Bock, G. R. and Goode, J. A. (ed.) (1998) *The Limits of Reductionism in Biology*. London: Wiley.

Collins, F. H. (1999) 'Shattock Lecture. Medical and societal consequences of the Human Genome Project', *New England Journal of Medicine*, 341, pp. 28–37. doi:10.1056/NEJM199907013410106.

Coyne, J. A. (2009) *Why Evolution is True*. Oxford: Oxford University Press.

Coyne, J. A. (2014) 'What scientific idea is ready for retirement?', https://www.edge.org/response-detail/25381. Retrieved 12 October 2020.

Darwin, C. (1868) *The Variation of Plants and Animals under Domestication*. London: Murray.

Darwin, C. (1871) *The Descent of Man: Selection in Relation to Sex*. London: Murray. 2004. Penguin Classics Reprint, with introduction by Moore, J. and Desmond, A.

Dawkins, R. D. (1976) *The Selfish Gene*. Oxford: Oxford University Press.

Dawkins, R. D. (2016) *The Selfish Gene. 40th Anniversary Edition*. Oxford: Oxford University Press.

Dawkins, R. D. (2019) 'Foreword', In Williams, G. C. *Adaptation and Natural Selection Princeton Science Library Edition*.

Dawkins, R. D. (2024) *The Genetic Book of the Dead*. London: Head of Zeus.

Deck, C., Jauker, M. and Richert, C. (2011) 'Efficient enzyme-free copying of all four nucleobases templated by immobilized RNA', *Nature Chemistry*, 3(8), pp. 603–608. doi:10.1038/nchem.1086.

Desmond, A. and Moore, J. (1991) *Darwin*. London: Michael Joseph.

DiFrancesco, D. and Camm, J. A. (2004) 'Heart rate lowering by specific and selective If current inhibition with ivabradine: A new therapeutic perspective in cardiovascular disease', *Drugs*, 64, pp. 1757–1765. doi:10.2165/00003495-200464160-00003.

Edelman, G. M. (1978) *Neural Darwinism: The Theory of Neuronal Group Selection*. New York: Basic Books.

Edelman, G. M., Gally, J. A. and Baars, B. J. (2011) 'Biology of consciousness', *Frontiers in Psychology.* doi:10.3389/fpsyg.2011.00004.

Felin, T., Koenderink, J., Krueger, J. I., Noble, D. and Ellis, G. F. R. (2021) 'The data-hypothesis relationship', *Genome Biology,* 22, 57. doi:10.1186/s13059-021-02276-4.

Futuyma, D. and Kirkpatrick, M. (2018) *Evolution.* Oxford: Oxford University Press.

Gluckman, P. D. and Hanson, M. A. (2005) *The Fetal Matrix: Evolution, Development and Disease.* Cambridge: Cambridge University Press.

Gluckman, P. D. and Hanson, M. A. (2006) *Mismatch: Why our World No Longer Fits Our Bodies.* Oxford: Oxford University Press.

Grove, A. J. and Newell, G. E. (1945) *Animal Biology.* London: University Tutorial Press. 1945 Reprint of 1944 2nd Edition.

Hingorani, A. D., Gratton, J., Finan, C., *et al.* (2023) '*BMJMedicine*', 2, p. e000554. doi:10.1136/bmjmed-2023-000554.

Huxley, J. S. (1942) *Evolution. The Modern Synthesis.* London: Allen & Unwin.

Kenny, A. J. P. (1963) *Action, Emotion, and Will.* London: Routledge and Kegan Paul.

McLaren, A. and Michie, D. (1958) 'An effect of uterine environment upon skeletal morphology in the mouse', *Nature,* 181, pp. 1147–1148.

Natarajan, C., Hoffmann, F. G., Weber, R. E., Fago, A., Witt, C. C. and Storz, J. F. (2016) 'Predictable convergence in hemoglobin function has unpredictable molecular underpinnings', *Science,* 354, pp. 336–339. doi:10.1126/science.aaf9070.

Noble, D. (1967a) 'Charles Taylor on teleological explanation', *Analysis,* 27, pp. 96–103.

Noble, D. (1967b) 'The conceptualist view of Teleology', *Analysis,* 28, pp. 62–63.

Noble, D. (2011) 'Neo-Darwinism, the modern synthesis and selfish genes: Are they of use in physiology?', *Journal of Physiology,* 589, pp. 1007–1015. doi:10.1113/jphysiol.2010.201384.

Noble, D. (2016) *Dance to the Tune of Life. Biological Relativity.* Cambridge: Cambridge University Press.

Noble, D. (2017) 'Evolution viewed from physics, physiology and medicine', *Interface Focus.* doi:10.1098/rsfs.2016.0159.

Noble, D. (2019) 'Exosomes, gemmules, pangenesis and Darwin', In *Exosomes. A Clinical Compendium* (ed. Edelstein, Smithies, Quesenberry, Noble), pp. 487–501.

Noble, D. (2022) 'How the Hodgkin cycle became the principle of biological relativity', *Journal of Physiology,* 600, pp. 5171–5177. doi:10.1113/JP283193.

Noble, D. (2024) 'Genes are not the Blueprint for Life', *Nature,* 626, pp. 254–255.

Noble, D. and Hunter, P. J. (2020) 'How tolling genomics to physiology through epigenomics', *Epigenomics,* 12(4), pp. 285–287. doi:10.2217/epi-2020-0012.

Noble, D. and Noble, R. (2021a) 'The Origins and Demise of Selfish Gene Theory', *Theoretical Biology Forum*, 115, pp. 152–161. doi:10.19272/202211402003.

Noble, D. and Noble, R. (2021b) 'Rehabilitation of Karl Popper's Ideas on Evolutionary Biology and the Nature of Biological Science', In *Karl Popper's Science and Philosophy* (ed. Parusniková, Z. and Merritt, D.). Cham: Springer. doi:10.1007/978-3-030-67036-8_11.

Noble, D. and Noble, R. (2025) *Darwin, Consciousness, and the Physiology of Agency*, In Press.

Noble, D. and Phillips, D. (2024) 'Speciation by physiological selection of environmentally acquired traits', *The Journal of Physiology*, 602(11), pp. 2503–2510. doi:10.1113/JP285028.

Noble, R. and Noble, D. (2017) 'Was the Watchmaker Blind? Or was she one-eyed?', *MDPI Biology*, 6, pp. 47. doi:10.3390/biology6040047.

Noble, R. and Noble, D. (2018) 'Harnessing stochasticity: How do organisms make choices?', *Chaos*, 28, pp. 106309. doi:10.1063/1.5039668.

Noble, R. and Noble, D. (2019) 'A-Mergence of Biological Systems', In *The Routledge Handbook of Emergence* (ed. Gibb, S., Hendry, R. F. and Lancaster, T.). doi:10.4324/9781315675213.

Noble, R. and Noble, D. (2020) 'Can reasons and values influence action: How might intentional agency work physiologically?' *Journal for General Philosophy of Science*. doi:10.1007/s10838-020-09525-3.

Noble, R. and Noble, D. (2023) *Understanding Living Systems*. Cambridge: Cambridge University Press.

Noble, R., Tasaki, K., Noble, P. J. and Noble, D. (2019) 'Biological relativity requires circular causality but not symmetry of causation: So, where, what and when are the boundaries?', *Frontiers in Physiology*, 10. doi:10.3389/fphys.2019.00827.

Rodwell, C. and Aymé, S. (ed.) (2014) *Report on the State of the Art of Rare Disease Activities in Europe*. Accessed March 7, 2021. http://download2.eurordis.org.s3.amazonaws.com/moca/other/2014%20Report%20on%20the%20State%20of%20the%20Art%20of%20RD%20Activities%20in%20Europe.pdf.

Romanes, G. (1886) *Post-Darwiniian questions. Isolation and Physiological Selection. Volume 3 of Darwin and After Darwin*. Chicago: Open Court Publishing.

Schulman, R., Yurke, B. and Winfree, E. (2012) 'Robust self-replication of combinatorial information via crystal growth and scission', *PNAS*, 109, pp. 6405–6410. doi:10.1073/pnas.1117813109.

Sun, Y. H., Chen, S. P., Wang, Y. P., Hu, W. and Zhu, Z. Y. (2005) 'Cytoplasmic Impact on Cross-Genus Cloned Fish Derived from Transgenic Common Carp (Cyprinus carpio) Nuclei and Goldfish (Carassius auratus) Enucleated Eggs', *Biology of Reproduction*, 72, pp. 510–515.

Taylor, C. (1964) *The Explanation of Behaviour*. London: Routledge Kegan Paul.

Taylor, C. (1967) 'Teleological explanation — a reply to Denis Noble', *Analysis*, 27, pp. 141–143.

Tinbergen, N. (1963) 'On aims and methods of ethology', *Zeitschrift für Tierpsychologie*, 20, pp. 410–433.

Williams, G. C. (1966) *Adaptation and Natural Selection. A Critique of some current evolutionary thought*. Princeton: Princeton University Press.

Wolpert, L. (1998) In Bock & Goode, 1998, p. 221.

Yanai, I. and Lercher, M. (2020) 'A hypothesis is a liability', *Genome Biology*, 21, pp. 231. doi:10.1186/s13059-020-02133-w.

Postscript: The Artist Disappears?*

With a standing ovation still ringing in the ears of us six musicians as we walked out, I was dragged back into the hall where a group from the audience wanted photographs. I smiled as best I could. Someone earlier in the interval had said that I looked very pale. I was certainly very tired, but happy tired – the kind that comes from the joy of performing really well. In fact, we had just agreed amongst the performers that we had performed our best concert ever. The percussionist, Keith Fairbairn,[1] had revealed himself as a melodic magician. Bryan Vaughan had performed with great feeling and range as the evening's anchor. Ray,[2] my brother, had excelled himself – the audience had been swinging in their seats as he thrilled them with one scintillating song after another with an astonishing range: one moment he was a quiet troubadour singing Jaufre Rudel's (1180)[3] song about distant love (*amor de terra lonhdana*), having seamlessly transformed himself from a modern rock star, dance and all. The last song I had sung in the language of the Troubadours ends 'I know stars that will turn your head' (*coneissi las estalas qui hen virat los caps*).

As I looked up at the back of the hall to smile for the camera, my head started turning. My quick instinct was very fortunate – I sat down immediately. If I had waited any longer – the cameras were still clicking – I would have fallen with goodness knows what consequences if that

---

*The title of Chapter 10 of *The Music of Life*.
[1] http://www.keithfairbairn.co.uk/.
[2] http://www.raynoble.com/index.html.
[3] http://en.wikipedia.org/wiki/Jaufre_Rudel.

Figure P.1.   The Holywell Music Room in Oxford is the oldest (1742) purpose-built concert hall in Europe. It was used by Handel and Haydn. To perform in it is a joy, the acoustics are so good, particularly so to a full and enthusiastic audience on Tuesday night, 17th August 2010.

spinning head had crashed hard into the polished wooden floor of the concert hall. But even that rapid precaution was insufficient. As friends rushed to help and to hold me up, I completely blacked out (Figure P.1).

Before I knew much about what was happening, I was being rushed in the emergency ambulance with an oxygen mask, beeping ECG machine … the complete works.… The intensive care unit at the John Radcliffe Hospital appeared later as a stellar sky of blinking lights and great hush. It was obvious what people suspected had happened.

In fact, as I came round more fully and greeted my brother ('You were bloody brilliant tonight!' – he responded, 'You stole the show!'), I knew that, most likely, that could not be the case. You can't be a cardiac physiologist for half a century without some degree of clinical instinct creeping in. But for two days, the medical team pursued this line to the limit, as they should have done, of course. They also thought they were treating a concert guitarist (the tell-tale fingernails) just suffering a 'crise cardiaque' at the end of an exhausting performance.

Knowing that it was quite unlikely to be a heart problem, I was almost unreasonably humorous, so much so that the alarm bells kept ringing on the blinking light machines each time I laughed. The medical team must have wondered what kind of joker they had just admitted to their normally eerily quiet sanctum.

Particularly when treated by cardiologists, I am more than happy to be incognito. That they thought they were simply treating a musician suited me perfectly. But that artist's mask came off when a young doctor who had attended lectures in my department 10 years previously appeared in my room: 'You must be the Denis Noble who wrote that fabulous book, *The Music of Life!*' Nice compliment, of course, but the basic puzzle remained. An older doctor summed it up: 'People don't just black out from a sitting position'. They had in fact recorded a substantial difference between my standing and sitting blood pressures. Sitting down should have restored the pressure to a safer level. Yet, the home care specialist was already working out how I could be cared for on a long-term basis after being sent home. Fortunately, the consultant was cautious and kept me in for a further day.

The resolution came the first time I went to the bathroom myself. The swimming head returned, and I was reduced to being on all fours on the floor. I was quick to pull the red alarm cord. They were even quicker with the oxygen and ECG recorders. They had the opportunity they needed: to observe the event in detail at close quarters. The next morning, I was in the Endoscopy Unit, under general anaesthesia, with a complicated apparatus down my throat, to be treated successfully for a duodenal ulcer from which, clearly, copious bleeding had occurred. QED.

At least I now know that my heart is in pretty good shape and, of course, that it wasn't a stroke either.

But I can't help asking some further physiological questions, because this episode, which took me so close to a final disappearing act, is in fact a neat example of the systems approach in a clinical context. When precisely did the trigger catastrophe actually happen? It must have happened before the concert started. My daughter had noticed that I looked tired. So did someone else who knew me well, in the interval when I took a walk in the fresh air to 'cool down' the sweating I attributed to exhaustion – I was the lead performer for most of the first half of the concert.

So, how could someone suffering an event that would produce severe hypotension continue right through a two-hour concert before succumbing to the inevitable collapse? The answer is in the small molecule we call adrenaline. Fired up as *The Oxford Trobadors*[4] always are before and during a performance, that little molecule, pouring out of my adrenal glands in response to my nervous signals, kept my blood pressure just above the lowest level and for just long enough to complete the performance, even including a magical encore after I had triumphantly called each of the musicians in turn to receive their deserved accolades from the audience.[5]

Some of the mechanisms of adrenaline's actions in accelerating heart rhythm and so maintaining blood pressure were worked out in my own laboratory (Brown *et al.*, 1975, 1978, 1980; Cohen *et al.*, 1978; Egan *et al.*, 1987, 1988; Hauswirth *et al.*, 1968, 1969; McNaughton and Noble, 1973; Noble, 1975a, 1975b). The approach to a sudden collapse of the circulatory system is also well analysed in systems physiology (Joyner, 2009), and the suddenness of the final event, when adrenaline can no longer maintain the system, is a subject of further investigation.

The standing ovation occurred after this encore when we sang in Korean a song, Mannam (만남), that is extremely popular in Korea. It was magical because, as soon as Ray sang the first line,[6] a 'choir' of 40 people emerged gradually in a crescendo from within the audience itself. A group of teachers from Kwangju (光州 광주) in South Korea were distributed all through the audience. They started standing and swaying as the song

---

[4] http://www.musicoflife.co.uk/music.html.

[5] The same mechanism must have been working for Kathleen Ferrier when she bravely performed her final role in Gluck's *Orfeo ed Euridice* at Covent Garden Opera House in 1953, despite being in great pain. She also left the performance on a stretcher. Soldiers in battle also know the phenomenon, when they can continue despite severe injury until the adrenaline surge fades and the inevitable collapse follows. Fortunately, my situation was different in all other respects. There was no pain. Just the sudden blackout.

[6] 우리 만남은 우연이 아니야 : Our meeting, it was not by chance.

Figure P.2. Ray Noble (centre) leading the performance of *Los de qui cau* at the *Oxford Trobadors* Concert in the Holywell Music Room. This song was composed by *Nadau* (see Chapter 9) and is one of the most beautiful and popular of the modern Occitan songs.

reached its climax and Ray outperformed himself by going up a whole octave to rise above the massed singing (Figure P.2).

The collapse afterwards was, of course, all the more dramatic. That is why sitting down was not sufficient to contain the problem.

The artist disappears? Well, no – not yet anyway.

References

Brown, H. F., McNaughton, P. A., Noble, D. and Noble, S. J. (1975) 'Adrenergic control of cardiac pacemaker currents', *Philosophical Transactions of the Royal Society B*, 270, pp. 527–537.

Brown, H. F., Noble, D. and Noble, S. J. (1978) 'The initiation of the heartbeat and its control by autonomic transmitters', In *Developments in Cardiovascular Medicine* (ed. Dickinson and Marks), pp. 31–52.

Brown, H. F., Noble, D. and Noble, S. J. (1980) 'Le rythme cardiaque: Les mécanismes ioniques de son contrôle sous l'influence de l'adrenaline et de l'acetylcholine', In *La transmission neuromusculaire. Les médiateurs et le "milieu interieur"*, pp. 207–229. Paris, New York, Barcelona, Milan: Masson.

Cohen, I., Eisner, D. A. and Noble, D. (1978) 'The action of adrenaline on pacemaker activity in cardiac Purkinje fibres', *Journal of Physiology*, 280, pp. 155–168.

Egan, T., Noble, D., Noble, S. J., Powell, T. and Twist, V. W. (1987) 'An isoprenaline activated sodium-dependent inward current in ventricular myocytes', *Nature*, 328, pp. 634–637.

Egan, T., Noble, D., Noble, S. J., Powell, T., Twist, V. W. and Yamaoka, K. (1988) 'On the mechanism of isoprenaline- and forskolin-induced depolarization of single guinea-pig ventricular myocytes', *Journal of Physiology*, 400, pp. 299–320.

Hauswirth, O., Noble, D. and Tsien, R. W. (1968) 'Adrenaline: Mechanism of action on the pacemaker potential in cardiac Purkinje fibres', *Science*, 162, pp. 916–917.

Hauswirth, O., Noble, D. and Tsien, R. W. (1969) 'Reconstruction of the actions of adrenaline and calcium on cardiac pacemaker potentials', *Journal of Physiology*, 204, pp. 126–128.

Joyner, M. J. (2009) 'Orthostatic stress, haemorrhage and a bankrupt cardiovascular system', *Journal of Physiology*, 587, pp. 5015–5016.

McNaughton, P. A. and Noble, D. (1973) 'The role of intracellular calcium ion concentration in mediating the adrenaline-induced acceleration of the cardiac pacemaker potential', *Journal of Physiology*, 234, pp. 53–54.

Noble, D. (1975a) 'Actions of catecholamines on ionic currents in cardiac muscle', In *Contraction and Relaxation in the Myocardium* (ed. W. G. Nayler), pp. 267–291. Academic Press.

Noble, D. (1975b) *The Initiation of the Heartbeat*. Oxford: Oxford University Press.

Glossary of Electrophysiology

In each entry, cross-references to other items in the glossary are indicated in bold type.

Action potential
Nerves, muscles and other excitable cells communicate by sending electrical signals, called action potentials, along their fibres. The signal consists of a transient reversal of the normal membrane potential, first with a phase of **depolarisation** that reverses the normal negative potential to become positive and then a phase of **repolarisation** that restores the internal negative potential.

Activation curve
Ion channels that are **gated** by voltage do not all switch on at the same voltage. There is a gradual change from all channels closed to all channels open. The curve relating the number of open channels to the membrane potential is called the activation curve.

Activation energy
A chemical reaction from state A to state B occurs at a speed that is determined by how much energy is required to make the change. When the energy is large, the speed is slow. You can think of this like a mountain between the two valleys, A and B. A high mountain (high activation energy) is more difficult to cross than a low mountain. Activation energies also determine how temperature influences the speed. This is also sometimes expressed as a Q_{10}.

Allele
The precise DNA sequence for any particular **gene** can vary from individual to individual. These variants are called alleles. In a population, there can be many alleles. Since most organisms have two sets of chromosomes, each individual has two copies of each **gene**, which may be the same or different alleles.

Atrium
The chambers into which blood flows into the heart are called atria. They then pass the blood into the **ventricles** of the heart.

Background sodium current
Not all **ion channels** that play a major role in the heart are gated by voltage. An important example is the channel that conducts what we call the background sodium channel, i_{bNa}. This was originally a prediction from the modelling of **sinus node** pacemaker activity. We had to postulate the existence of a channel conducting sodium ions all the time in order for the modelling to work. Later, work in Irisawa's laboratory in Japan and our own laboratory succeeded in characterising its ion selectivity and other properties.

Bessel functions
These are mathematical functions that form solutions to certain differential equations applied to cylindrical or spherical co-ordinates. There are various kinds, including Bessel functions using imaginary arguments, which is the kind referred to in Chapter 1.

Cable theory
The equations of cable theory were originally developed for the flow of electric current in metal cables, such as the first transatlantic cables. They were applied to nerve axons since these also function like cables. The most extensive treatment of cable theory applied to biological systems is *Electric Current Flow in Excitable Cells* (Jack, Noble and Tsien, 1975).

Calcium-induced calcium release
Calcium ions are stored inside muscle cells in the **sarcoplasmic reticulum**. The trigger for releasing the stored calcium to activate contraction is calcium itself. So, calcium ions enter the cell through calcium **ion**

channels and then cause the release of a much larger quantity of calcium inside the cell. This is called calcium-induced calcium release.

Conductance
The transport of ions through **ion channels** forms an electric current. The magnitude of the current depends on the channel conductance. The higher the conductance, the more current flows. An alternative measure of the speed of transport is the **permeability**, which can also apply to substances that are not charged.

Depolarisation
Cells are polarised. The interior of the cell is usually negative. During excitation, entry of cations, which are positively charged, reduces the polarisation. This process is called depolarisation. It was originally thought that this process only went as far as removing the negative potential. In fact, it reverses it so that the cell becomes positive. But we still refer to this as depolarisation. The reverse process is called **repolarisation**.

Electrocardiogram
The changes in electrical potentials inside cardiac cells during each beat of the heart create electric currents that flow between different parts of the heart. A small fraction of these currents can be detected at the surface of the body. These recordings are called electrocardiograms. Doctors use them routinely to check on heart activity and to help in the diagnosis of disease states.

Electrogenic
This refers to any structure or mechanism that carries an electric current. **Ion channels** are obvious examples since they form pores in the cell membrane through which the charged ions can flow. Proteins that exchange ions, such as sodium–potassium exchange (the **sodium pump**) or the sodium–calcium exchanger, are also electrogenic because the movement of charge in the two directions does not balance. The sodium pump transports 3 sodium ions in exchange for 2 potassium ions, so it carries a net positive charge out of the cell. The **sodium–calcium exchanger** was originally thought to be electrically neutral (2 sodium ions in exchange for one divalent calcium ion). In fact, it has been found to transport an extra sodium ion (so the ratio is 3:1) and is therefore electrogenic.

Entropy of activation

The activation energy of an ion channel **gating** reaction with high temperature dependence can be so large that, if there were nothing else to favour the reaction, it would not proceed in a biologically relevant time scale. This is the case for the **pacemaker** current, i_f, with a Q_{10} of 6 (Chapter 2). One possible explanation is that in the formation of the activated complex, there is an entropy change favouring the reaction. A different possibility is that we are dealing with a chain of reactions, with the temperature dependence of one reaction changing the conditions for a second reaction.

Equilibrium potential

See **Reversal potential**.

Error functions

Error functions are so called because they are sigmoid functions that were developed for use in probability and statistics. They have also proved very useful in solving partial differential equations, which is their application in **cable theory**.

Excitation threshold

Action potentials in excitable cells, like nerves and muscles, are 'all or nothing'. Once the cell has been excited to initiate the action potential, it continues to generate itself using positive feedback between the membrane voltage and activation of the **ion channels**. For example, once sufficient sodium channel **gates** have been opened, sodium ions flow into the cell in sufficient quantity to depolarise the membrane further, which opens even more channels. This process continues until the membrane potential approaches the **reversal potential** at which the sodium chemical and electrical gradients balance each other.

Gap junction

See **Nexus** junctions.

Gating

Ion channels are formed by proteins that sit in the cell membrane and form a pore through which ions can cross the membrane. Most channels are gated. Part of the protein itself, or another subunit or chemical (sometimes, ions themselves can be gates), moves to close or open the channel.

In excitable cells, the main channel proteins involved are gated by the membrane potential. The gating part or molecule moves in the electric field because it is charged. The best documented example is the sodium channel inactivation gate, which consists of a charged section of the amino acid chain that connects two of the transmembrane domains (III and IV) and which swings in the electric field to close the channel by sitting at its intracellular opening. This is the process of **inactivation**.

Gene
The modern molecular biological definition of a gene is a section of DNA sequence forming a template for the production of a protein or RNA. The original definition of a gene, introduced a century ago, was the inheritable cause of a particular phenotype. The cause of any particular phenotype, however, extends way beyond any particular DNA sequence. The two definitions of a gene are not therefore equivalent. This is the cause of much confusion in the literature of genetics, both popular and scientific. For further development of this critique of the concept of a gene, see Chapter 9.

Giant patch clamp
This is a development of the **patch clamp** to enable large areas of cell membrane to be controlled to enable ionic currents to be resolved that are not necessarily generated by ion channels. It was developed by Don Hilgemann (see Chapter 5) and used by him to great effect in studying the **sodium–calcium exchanger.**

Hodgkin–Huxley equations
These are differential equations formulated by Alan Hodgkin and Andrew Huxley in 1952 to describe the **gating** kinetics of the sodium and potassium **ion channels** in the giant nerve fibre of the squid. The equations are based on the gating reaction being the movement of a charged particle or part of the channel in the membrane electric field.

Holding potential
In a **voltage clamp** experiment, the membrane potential is usually held constant at a level called the holding potential between each pulse. In Hodgkin and Huxley's original work, the holding potential was chosen to be the resting level of the membrane potential, but this is not necessarily required. In a spontaneously beating cell, such as a cardiac pacemaker

cell, there is no resting potential anyway. The holding potential is then chosen for convenience, usually outside the range of the **activation curve** for the **ion channel** being studied.

Hyperpolarisation
This is the opposite of **depolarisation**. Instead of reducing the negative intracellular potential, it increases it. It was through hyperpolarising the membrane of sinus node cells that Brown, DiFrancesco and Noble (1979) discovered the 'funny' current, i_f. This is sometimes called a hyperpolarising-activated **ion channel** to distinguish it from channels that are activated by depolarisation.

Inactivation
In addition to displaying an **activation curve**, many **ion channels** with **gating** by voltage also display inactivation so that the ionic current flowing through the channel is transient. The first example of this process was the sodium channel investigated by Hodgkin and Huxley, which inactivates quite rapidly following the activation process. The transient outward current referred to in Chapter 2 is an example of a potassium channel that displays inactivation.

Ion channels
Since ions are charged, they do not pass through the lipid bilayer forming the cell membrane. The lipids repel charged ions, just as they also repel polar molecules like water. Ion channels are pores that allow ions to cross the membrane without having to pass through the lipid layer. The pores are formed by proteins. There are ion channels that selectively pass sodium, calcium, potassium or chloride ions, and channels that allow protons (hydrogen ions) to pass.

Matrices
A matrix in mathematics is an array of numbers arranged in rows and columns. These could correspond, for example, to the coefficients of interaction between the components of a system. Matrices are important in algebra and they can also be of great use in solving systems of partial differential equations of the kind often used in electrophysiology.

Nernst equation
This is the equation for the **reversal potential** (equilibrium potential) of an **ion channel** selective for a single ion. In the absence of an electric

field, the gradient causing flow of ions through the channel will be the chemical gradient, which exists when the ions are more concentrated on one side of the membrane than on the other. An electric field can be set up that opposes this flow. For example, potassium ions flow out the cell under the influence of the chemical (concentration) gradient, but they can be held in the cell by the negative electrical potential of the interior. The potential at which the two forces exactly balance, and at which there is no net movement, is called the equilibrium potential. The Nernst equation is the equation for this balance. In the case of a single charged (monovalent) ion, the equation predicts a roughly 60mV change in the equilibrium potential for a tenfold change in concentration. This therefore became one of the tests for the ionic selectivity of a channel.

Nexus
This is a connection between two cells formed by proteins called connexins.

Nonlinearity
In mathematics, we distinguish between linear and nonlinear functions. For example, the simplest model of an **ion channel** would be one in which the flow of current is proportional to the chemical and electrical gradients (see **reversal potential**). Very often, this is not true and the relationship is nonlinear. Also, a channel gated by voltage might be linear (or close to linear) for sudden changes, but nonlinear once the gating process occurs. Linear systems are easier to solve mathematically, but as the work in Chapter 3 shows, even some highly nonlinear systems can be solved with closed-form solutions.

Pacemaker
The heartbeat is a nearly regular rhythm, approximately 60 beats per minute in a human. The areas of the heart that generate rhythm are called pacemakers. The natural pacemaker is the **sinus node**, a small region of the heart situated where the veins empty blood into the **atria**. There is also a region joining the **atria** and **ventricles,** which is called the atrio-ventricular node, which also generates rhythm. The **Purkinje fibres** can also display rhythm. Normally, the sinus node is the fastest pacemaker and so excites the heart before the other regions can do so. An ectopic pacemaker is one that arises because of a disease state (such as that described in Chapters 3 and 5) in a region that does not normally display rhythm. This can then compete with the natural pacemaker to produce arrhythmia.

Pacemaker depolarisation

Excitation of a nerve or muscle cell is achieved by depolarising it to the **voltage threshold**. In **pacemaker** regions, the depolarisation occurs spontaneously as a gradual process called pacemaker depolarisation. A major goal of cardiac electrophysiology is to work out the **ion channel** mechanisms generating this depolarisation.

Patch clamp

The earliest techniques for controlling the membrane potential (**voltage clamp**) used large electrodes that could be inserted into very large nerve fibres, like those found in the squid. Work on the small cells of the heart became possible because of the invention of glass microelectrodes with tips around 1 μm in size. These can penetrate the cell membrane, which then reseals around the glass to enable the pipette to be used to record voltage or to inject current. A major advance came when Erwin Neher and Bert Sakmann discovered that if the pipette tip was made smooth enough, it was possible for the glass pipette to seal to the membrane without penetrating it. This allowed the experimenter to study the ion channels in the patch of membrane sealed to the electrode. Alternatively, the patch could be sucked up to break it and enable the patch electrode to record or inject current for the cell as a whole. A further development of this method involved using much larger electrodes – the **giant patch clamp**.

Permeability

This is a measure of the speed with which a substance can move through something like a membrane. **Ion channels** are what enable cell membranes to be permeable to ions. The permeability of an **ion channel** determines its electrical **conductance**: for the movement of charged items, conductance is an alternative way of expressing permeability, though the two are not defined in the same way.

Perturbation theory

This involves mathematical methods that are used to find an approximate solution to a problem that cannot be solved exactly, by starting from the exact solution of a related problem. Perturbation theory is applied by adding a small term to the mathematical description of the exactly solvable problem. The theory then leads to an expression for the desired solution in terms of a power series in a small parameter that quantifies the deviation from the exactly solvable problem. This is the method that was used with Dario DiFrancesco in Chapter 5 to resolve problems arising from ion accumulation.

Physiome Project
This project to model the cells, organs and systems of the body is described in Chapter 7. It was launched in 1997 by the International Union of Physiological Sciences (IUPS).

Plateau phase
Depolarisation in most heart cells is very fast, just as in nerve and skeletal muscle. **Repolarisation**, however, is very slow. The cell remains depolarised for a much longer period of time. This produces a very long **action potential**. The slowest phase of repolarisation is called the plateau.

Polynomials
Polynomial functions are mathematical series of finite length consisting of terms using whole-number exponents and which can also include addition, subtraction and multiplication. Their use in electrophysiology is described in Chapter 3.

Purkinje fibres
In the hearts of large animals, the speed of activation of the **ventricle** would be too slow to ensure synchronous contraction if the conduction occurred only through the ventricular muscle itself. A fast-conducting network of fibres has developed to ensure synchrony. This network was first identified in 1839 by the Czech physiologist Jan Purkyně. The speed is achieved by the cells being very large (see Chapter 3 for the relevant equation). This is why they were chosen for the early **voltage clamp** experiments. It was easier to penetrate large cells with two microelectrodes. All the experimental work for the Noble 1962, McAllister-Noble-Tsien 1975 and DiFrancesco-Noble 1985 models was done on Purkinje fibres.

Q_{10}
A standard way to determine the **activation energy** for a chemical reaction is to measure the way in which its speed varies with temperature. At higher temperatures, more molecules have sufficient energy to reach the activation energy level. If the activation energy is very high, the

temperature dependence will also be high. In biological systems, because the temperature range that can be used is fairly limited, it has become customary to measure the ratio of the speeds at two temperatures 10°C apart. This ratio is called the Q_{10}. For the **gating** reactions of most **ion channels**, this ratio is in the range 2-3 which corresponds to reasonable activation energies. A surprise in the work described in Chapter 2 was that the slowly activated current, the i_f channel, in the **pacemaker** range of potentials has an unusually high Q_{10} of 6.

Repolarisation
This is the process by which the initial **depolarisation** of the **action potential** is reversed. The process is fragile in the heart, which is why many disease states and drugs cause arrhythmias.

Resistance
This is the inverse of conductance. A high conductance means a low resistance and vice versa.

Reversal potential
This is the potential at which an ion channel current reverses direction. In the case of a channel conducting just one ion species, e.g. potassium, this potential would be equal to the equilibrium potential given by the **Nernst equation**. When more than one ion species is carried, the reversal potential is not equal to either of the equilibrium potentials. There are equations for this situation, referred to in some of the papers, but they are not discussed in the chapters of this book.

Sarcoplasmic reticulum
This is a network (hence reticulum) in the muscle cytoplasm (hence sarcoplasmic) that stores calcium ions. The release of calcium from these stores is achieved by the **calcium-induced calcium release** process.

Sinus node
This is the normal pacemaker region of the heart, found in the region where the great veins empty into the atrium.

Slope conductance
Membrane **conductance** can be measured by applying small voltage deflections and measuring the current that flows. This might be equivalent

to the net membrane conductance, but this is not necessarily true since there may be **gating** reactions occurring sufficiently rapidly to react to the voltage change before the current change is measured. The conductance that is then measured is a tangent (slope) to the relevant current–voltage relation of the cell. The slope conductance can even change in the opposite direction to the net conductance, as described in Chapter 1.

Sodium–calcium exchanger

This is the protein that uses the sodium gradient to push calcium ions across the cell membrane. Discovered in the heart by Harald Reuter, it was originally thought to be electrically neutral, transporting two univalent sodium ions in exchange for each divalent calcium ion, called a 2:1 stoichiometry. It is now known to have a stoichiometry of 3:1 (see Chapter 5) and is therefore **electrogenic**.

Sodium pump

This is a protein that forms an exchanger for sodium and potassium. Three sodium ions are pumped out of the cell in exchange for two potassium ions. The mechanism is therefore **electrogenic**. It is often called the Na^+-K^+ATPase since it transports the ions by splitting ATP.

Sodium threshold

This is the membrane potential at which the sodium channel current becomes sufficiently large to initiate an all-or-nothing **action potential**.

T wave

This is the wave of the electrocardiogram (Chapter 4) that corresponds to the **repolarisation** of the **ventricles**.

Transfer function

In the Tsien and Noble (1969) analysis of ion channel kinetics using **transition state theory**, this corresponds to the ion transport of a channel protein in the absence of **gating**. If the channel were a simple conductance, this function would be linear. In fact, it often displays **nonlinearity**.

Transition state theory

This is the chemical theory of how **activation energies** determine the rate of a chemical reaction.

Ventricle
The ventricles are the thick muscles of the heart that pump blood at high pressure into the aorta.

Voltage clamp
This is a technique for keeping the membrane potential at values chosen by the experimenter. Since voltage is what controls the **gating** of **ion channels**, it is important to control this parameter.

Voltage-dependent gating
This refers to the opening and closing of ion channels by gates that move in the electric field. The reactions described in the Hodgkin–Huxley nerve equations are voltage-dependent gating reactions.

Voltage threshold
This is the voltage at which an all-or-nothing action potential or repolarisation is triggered.

Index

√–1, 16
1962 model, 13, 16–17, 25, 29

A
acetylcholine, 87
Action, Emotion and Will, 162
action potential duration, 55
activation energy, 28
adrenaline, 34, 66
agnostics, 146
Algol, 81
Alighieri, Dante, 68
allele, 97
all or nothing, 42
all-or-nothing repolarisation, 32
analysis, 130
Analytical Approaches Using
    Polynomial Models, 41
analytical approaches with closed-
    form solutions, 152
analytical mathematics, 112, 153
analytical solution, 39–40, 45
Annual Review Lecture, 73
Aristotle, 143, 171
association scores, 104
atheism, 146
atrial action potential, 84

atrial cell model, 85
atrial fibrillation, 57
atrial model, 81
atrium, 81
attractors, 153
Auckland, 110, 112–114, 116
Auckland University, 117
Austin, J. L., 22
autocode, 2
A-V block, 57
Ayer, A. J., 22

B
Babbage, Charles, 52
Bacaner, Marvin, 109
background sodium channel, 103
background sodium current channel,
    102
backup mechanisms, 102, 105
Balliol College, 83, 130
Balliol Interdisciplinary Institute
    (BII), 131
Banister, Jean, 85, 87
Basel, 137
Bassingthwaighte, Jim, 115, 120
Batchelor, Stephen, 147
Beeler–Reuter model, 35

US Army High-Performance
    Computer, 109
use and disuse processes, 169

V
valve computer, 2
venture capital, 116
virtual, 123
virtual human, 120
Virtual Physiological Human, 115,
    119, 155
virtual ventricle, 112
voltage clamped, 64
voltage clamp technique, 25
voltage threshold, 42

W
Wallace, Alfred Russel, 169
Waller, A. D., 54
Watergate Hotel, 57
Watson, James, 163
Watson, Thomas, 2
Weidmann, Silvio, 12–14, 33–34
Weismann, August, 169

Wellcome Trust, 80, 88
Westerhoff, Hans, 120
whole-cell mode, 83
whole-organ simulation, 114
Why Evolution is True, 163
Wilkes, Kathy, 130
Wilkie, D. R., 124
Williams, Bernard, 22
Williams, George, 162
Winslow, Rai, 58, 109, 114–115, 120
Wittgenstein's Tractatus, 141
Wolf, Naomi, 113
Wolpert, Lewis, 163
Women in Physiology, 85
World Congress, 114
Wuhan Fish Institute, 175

X
XML, 116

Y
yeast, 99, 120
yeast genome, 99
Young, J. Z., 47, 86, 124, 161